Studies in Computational Intelligence 412

Editor-in-Chief

Prof. Janusz Kacprzyk
Systems Research Institute
Polish Academy of Sciences
ul. Newelska 6
01-447 Warsaw
Poland
E-mail: kacprzyk@ibspan.waw.pl

For further volumes:
http://www.springer.com/series/7092

Omar K. Hussain, Tharam S. Dillon,
Farookh K. Hussain, and Elizabeth J. Chang

Risk Assessment and Management in the Networked Economy

Springer

Authors
Dr. Omar K. Hussain
Digital Ecosystems and Business
Intelligence Institute
Curtin University
Perth
Western Australia
Australia

Prof. Tharam S. Dillon
Department of Computer Science and
Computer Engineering La Trobe
University Melbourne, Victoria
Australia

and

Digital Ecosystems and Business
Intelligence Institute Pty Ltd
Perth, Western Australia
Australia

Dr. Farookh K. Hussain
School of Software
University of Technology, Sydney
Broadway
New South Wales
Australia

Prof. Elizabeth J. Chang
Digital Ecosystems and Business
Intelligence Institute
Curtin University
Perth
Western Australia
Australia

ISSN 1860-949X e-ISSN 1860-9503
ISBN 978-3-642-28689-6 e-ISBN 978-3-642-28690-2
DOI 10.1007/978-3-642-28690-2
Springer Heidelberg New York Dordrecht London

Library of Congress Control Number: 2012933084

Printed on acid-free paper

Springer is part of Springer Science+Business Media (www.springer.com)

Foreword

In the context of ICT, risk has been historically considered from a project-management perspective; however, risk is an inherent property of every human endeavor. Security research has focused on the risk due to outsider attacks, but it is now recognized that a major source of risk in economic and social processes is due to dysfunctional behavior of participants to distributed processes. From a classic rational standpoint, dysfunctional behavior on the part of a participant to a distributed protocol takes place when cooperation delivers a lower payoff than non-cooperative behavior. In turn, payoff depends on process economics, which evolves over time.

Much research work has been done on analyzing the risk of client-server and peer-to-peer interactions either in the presence of purely rational actors or in the presence of irrational, malicious actors. In the first case, game-theoretical approaches have been proposed where incentives and penalties are used to prevent misbehavior of rational actors.

The second case has been studied by the distributed computing and computer security communities, which have proposed protocols to alleviate the effect of randomly malicious behavior. At first sight, those two models seem to fit the behavior of both individuals and organizations: typically, organizations will mostly act rationally, while individuals will sometimes act rationally, but some other times will decide to deviate from rational behavior without apparent reason (think of a disgruntled Web server administrator who, having being fired, just wants to harm the outcome of a business process).

Unfortunately, in the last few years, we have realized that neither risk modeling approach is able to achieve full understanding of risks linked to online transactions. Laws, regulation and contracts have traditionally provided powerful "post-hoc" risk alleviation and reduction measures; but these traditional techniques have become largely ineffective due to changes in the technological scenario.

Today, doing business online is no longer an exception, but the rule: the bulk of economic and social activities is composed of Web-based interactive processes, often carried out on the global Net. Interactive online processes involve transactions taking place between loosely connected parties, often totally or partially unknown to one another. Emerging technology paradigms like the Cloud and Software-as-a-Service (SaaS) may have somewhat belittled the importance of traditional integration problems like invocation paradigms, but have made online risk prediction more an issue than ever.

This book, authored by some of the leading researchers in the field, includes chapters that shed a new light over the domain. The book builds on the important idea that users' confidence in online services will be increasingly based on a sound, shared notion of dynamic transactional risk, rather then on today's heuristics. Such notion, as proposed in this book, captures the fundamental aspects of occasional online transactions, including time-dependence and possible information asymmetries.

The book's chapters, scientifically rigorous but easy to understand even for the non-specialist, lead the reader toward a full understanding of what transactional risk is, explaining how it can be identified, computationally assessed and managed.

Also, the book provides a clear view of the new, intriguing research directions that are being opened up by the notion of dynamic transactional risk. I believe that this book sets an important milestone in the risk modeling and prediction domain.

Ernesto Damiani
Department of Information Technology
Università degli Studi di Milano, Italy

Preface

Advances over the past 15 years in the area of Information Communication Technologies (ICT) enable the web to act as a binding medium for users to carry out commercial and non-commercial activities, irrespective of their geographical location. It has created a networked economy that is dynamic, self-organized, and goal specific. Users can utilize the different platforms, architectures, and devices which this medium provides to complete their tasks in an efficient, reliable and faster way. In order to take advantage of the ease with which business associations are facilitated over the web, users are inevitably adopting this medium in their lifestyle to carry out their business activities to achieve their desired outcomes in a more efficient, faster and reliable way. The user may be an individual agent, SME or a business which wants to achieve certain aims or desired outcomes. Various researchers have published scholarly work in the literature, detailing how a user can maximise his/her interaction experience through this medium. However, one factor which has been omitted to a large extent in the literature, but one which is important for the user to consider in order to maximize his/her interaction experience is the notion of *transactional risk.* This is in contrast to the widely discussed notion of trust and security while making decisions in relation to business activities over the web. Analysis of transactional risk in such activities is important as it allows the user to determine the potential loss as a result of the failure of the business activity; therefore, it is a critical factor that needs to be taken into account during the decision-making process. This book is an effort in such a direction to identify, assess, analyse, manage and record transactional risk in e-business networked economy.

In order to achieve the aforesaid objective, this book is divided into eleven chapters. A brief introduction of risk, its discussion in various real-world applications, the factors important for its understanding, its importance while engaging in decision-making in business activities, the classification and importance of risk as compared to trust and security, the different types of risks and the relationship between trust and transactional risk are discussed in **Chapter 1**. **Chapter 2** discusses transactional risk in detail. In this chapter, the different sub-categories of transactional risk, the types of transactional risk and its characteristics are discussed. Subsequently, the various types of business activities and the events and uncertainties of transactional risk in business activities are discussed before summarizing the issues associated with transactional risk analysis in the literature. Furthermore, the ISO standard on risk management is discussed. This chapter concludes by discussing the two different ways of approaching risk management in business activities.

In **Chapter 3,** the key concepts relating to transactional risk analysis are defined, followed by a comprehensive operational definition of transactional risk, taking into account the object of analysis in the domain of business activities. An approach for transactional risk analysis in business activities is presented, and the different phases in that approach are discussed. The different ways to initiate the process of transactional risk assessment are then discussed and finally, the notions of risk set, or the risk relationship formed as a result of a risk activity are defined and discussed. In **Chapter 4,** the transactional risk identification phase in business activities is presented and discussed. The process of formalizing the assessment criteria of the business activity, and then the importance or the significance of the corresponding assessment criterion and its role in the different phases of risk analysis is outlined and discussed. Subsequently, a mathematical model for ascertaining the significance of the assessment criteria of the business activity is presented. This chapter concludes by discussing the division of time into different phases to capture the dynamic nature of transactional risk while undertaking its assessment.

In Chapters 5-6, the process of transactional risk analysis in business activities is discussed. In **Chapter 5,** the focus is on measuring the sub-category of performance risk assessment in a business activity. The chapter starts with a discussion on the importance of having a unified measure for performance risk assessment, before proposing the FailureLevel and the Failure Scale. Subsequently, the different time scenarios in which performance risk in a business activity has to be ascertained and the importance of considering the time delay factor to capture the freshness status while undertaking performance risk assessment are discussed. The process of classifying the recommendations according to their credibility is then discussed. After this, the process of ascertaining the FailureLevel in different time phases of the business activity and then summarising the assessment to ascertain the performance risk of the business activity is explained in detail. In **Chapter 6,** the focus is on measuring the sub-category of financial risk, starting with the identification of the object of analysis at stake and the different types of events to be considered while undertaking financial risk assessment in business activities. The various key terms and techniques to identify the variability of financial resources at stake in the business activity are proposed before discussing in detail the mathematical techniques for ascertaining the financial risk in a business activity with examples.

In **Chapter 7,** the analysis of Chapter 5 and Chapter 6 is combined to ascertain the transactional risk in the business activity. This chapter starts by identifying the variability and uncertainty to be identified during transactional risk assessment before defining the scale on which it can be determined in the business activity. Subsequently, the importance of semantics and the different representations, namely numeric and linguistic representations of transactional risk, are discussed. Techniques and steps in determining each representation of transactional risk are explained in detail with examples. Once the severity level/s of transactional risk and their likelihood of occurrence in the business activity are ascertained, **Chapter 8** focuses on the transactional risk evaluation phase. This chapter begins by first discussing the importance of risk propensity or risk attitude while undertaking transactional risk evaluation and then defines the different risk propensity levels, which we consider in our approach, to evaluate transactional risk. The proposed approach for transactional risk evaluation is then explained in detail with examples and the analyses of results from the approach are summarized.

In **Chapter 9,** the focus is on transactional risk management in business activities. A case study of just-in-time or demand-driven production is considered and discussed w.r.t. how the previously proposed frameworks in the book can measure the adjusted transactional risk in the business activity as a result of the outcomes of the interaction specific risk treatment steps. In **Chapter 10,** the focus is on the transactional risk recording phase. The two stages to be completed in this phase, namely determining the non-commitment in the business activity and adjustment of recommending agent's credibility are discussed. The mathematical techniques for both these stages are explained in detail with examples.

Chapter 11 concludes the discussions in this book by discussing briefly the future research directions.

Acknowledgements

We acknowledge the unique and supportive research environment provided by Digital Ecosystems and Business Intelligence Institute (DEBII) that allowed us to conduct this research work. We acknowledge the Australian Research Council for the Linkage Grant LP0560346 entitled *Trusted environment for Virtual Collaboration* which laid the background and foundation of this work. We also express our deep gratitude and sincere thanks to Ms Michele Mooney and Ms Bruna Pomella for proof reading the book.

I appreciate and am grateful for the support of my parents. my wife and my lovely daughter for their constant encouragement throughout the process of completing this book. I thank my parents for the sacrifices they made in helping me to reach a stage where I am at present.

- *Omar*

I wish to express my appreciation to my parents for their help and support through the years.

- *Tharam*

I would thank my parents, wife, and my little sons for all their support and help.

- *Farookh*

I would like to thank my late mother for her inspirational up-bringing and support.

- *Elizabeth*

Contents

Chapter 1
Understanding Risk and Its Importance in Business Activities

1.1 Motivation

Nearly all of us have used the word *risk* in our conversations or have heard it in our daily life experience. To mention a few such situations: (i) you risk losing your driving license if you are caught drink driving; (ii) you risk your enrolment being cancelled if you are caught cheating; (iii) you risk losing your job if you don't improve your performance; (iv) your business is at risk if the current financial and economic situation does not improve; (v) you risk not reaching your job interview in Sydney on time on Monday if you leave from Perth by car on Friday. In all the above mentioned situations, risk is associated with an outcome that highlights *losing* a certain thing; for example losing a license, losing an enrolment, losing a job, losing a business, not reaching an interview on time. In all these examples, the *thing* which may be lost is considered an important one, hence the word 'losing' is associated with them.

Based on the above discussion, it can be said that risk, in general terms, highlights the occurrence of an outcome that tends towards experiencing a loss of a certain thing. Although the interpretation and representation in such an everyday use of the word *risk* is similar across various real-life applications, it has a more technical meaning in different human endeavors such as business, industry, and manufacturing. Understanding risk from the perspective of such endeavors is therefore important due to the different aspects at stake. Our focus in this book is to understand, quantify, measure, minimize or alleviate risk in business activities. To achieve this, we first need to understand what risk is, what its characteristics are and why is it important to quantify and measure it? In the next section, we briefly consider some typical real world situations and scenarios which will help us to answering these questions.

1.2 Discussion of Risk in Real-World Situations

In this section, we discuss risk in different real-world situations. It is important to understand that the real-world human endeavor activities discussed may require more than one activity to achieve their final objective. Such situations are quite

O.K. Hussain et al.: Risk Assessment & Management in the Networked Economy, SCI 412, pp. 1–29.
springerlink.com

common in the business scenarios discussed, where a collection of sub-activities form an activity or a collection of activities form a business association or an interaction.

1.2.1 Importance of Risk Assessment in Business Interactions

Business interactions are the backbone which drives the economy of the modern world. They are carried out with the aim of achieving certain specific business outcomes that are consequential for the progression, advancement and sustenance of the users involved. By the term 'business interaction', we do not just mean e-commerce interactions carried out on the internet, but any interaction that is carried out in the wider domain of business. It is likely that such interactions or business contracts will have financial resources at stake in order to achieve the desired results as their outcomes. But there might be various uncertainties involved which might result in the non- or partial achievement of the business contract. As shown in Figure 1.1, the occurrences of any such uncertainties pose a *risk* to the successful completion of the business activity that will have serious consequences to the users involved. In today's competitive business world, the aim of participants is to minimize any such risks while simultaneously achieving maximum returns and benefits. Hence, it is necessary to carry out an analysis of the risk to ensure the successful achievement of the business objectives. The importance of correct understanding, analysis and characterization of risk is demonstrated by the current credit crunch and economic crisis.

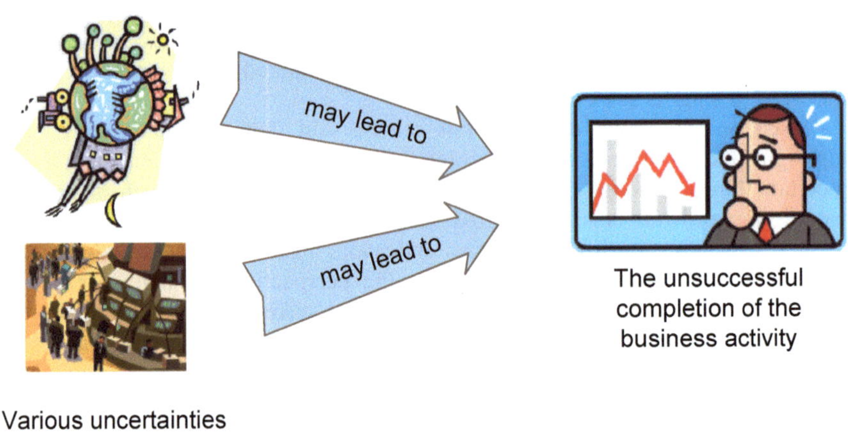

Fig. 1.1 Importance of risk assessment in business interactions

1.2.2 Importance of Risk Assessment in Demand-Driven Manufacturing

In today's competitive world, manufacturers are constantly subjected to massive pressures to reduce their operational costs and at the same time, improve or increase their production efficiency. Cost reduction implies that they have to shift or adopt a new process of producing or manufacturing their goods. To achieve this, demand-driven production is increasingly gaining popularity compared to batch production. Demand-driven production is based on real-time customers' orders and aims to drive demand planning accuracy and improve manufacturing efficiency. It frequently uses Just-in-Time (JIT) techniques with the basic strategy of matching the inputs required for the immediate production of goods, thus matching the exact orders and specifications of the customers, while maximizing value, eliminating waste and reducing the costs associated with inventory. Goods may be produced and assembled in the manufacturing company's own production plant, which is termed in-house production; or the production may be contracted to third party manufacturing companies, which is termed outsourced production; or it may use both these methods. Whatever production mechanisms are utilized, the basic requirement for the manufacturer is to ensure that the generated output from its production units matches the consumer's demands in that time period. But, there might be various uncertainties during this process which could result in the manufacturer not meeting the required demand [1]. Some of them to mention a few are:

a) The production units in the manufacturer's different plants may not operate according to their desired level of operation. This may be due to unexpected outages or planned maintenance downtime,
b) The required raw materials may not be available in the required quantity,
c) The required manpower for carrying out the required tasks is insufficient.

As shown in Figure 1.2, the occurrence of any of these uncertainties will pose a *risk* to the manufacturer in meeting the expected schedule and customers' demand over that given time period.

1.2.3 Importance of Risk Assessment in Power Generation

The deregulation of the power generation industry has created an open market which has led to fierce competition [2]. In order for a Genco to maintain an advantage over its competitors in such a market, it has to ensure that it meets the required energy demands of its customers. To do so, Gencos form bilateral contracts with their clients which guarantee their required energy. The bilateral contracts between the generation company and the clients are usually formed over a substantially large period of time. However, as these contracts are legally binding, the failure of a Genco to meet clients' needs will have various financial, economic and deleterious sustainable consequences to it. Figure 1.3 shows economic considerations and environmental factors vary constantly and there are various uncertainties which may result in *risk* of the Genco not meeting the required demand of the clients. So, it is imperative that Gencos proactively assess the risk which will ensure that they commit to their formed bilateral contracts.

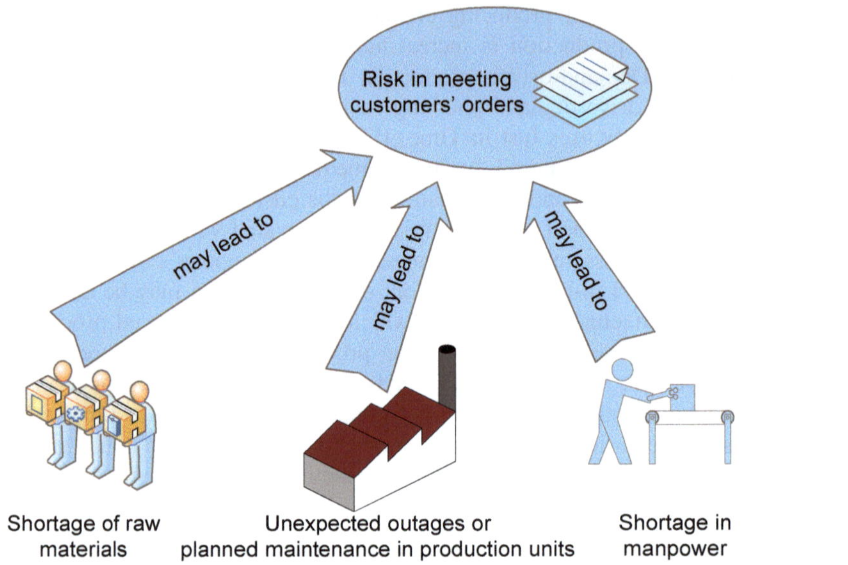

Fig. 1.2 Importance of risk assessment in demand-driven production

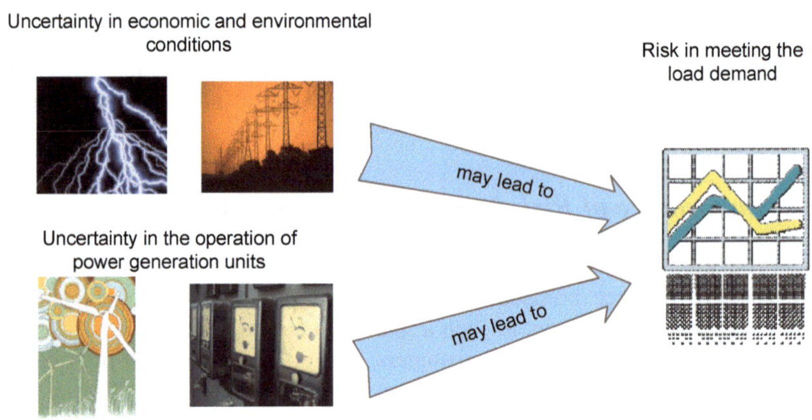

Fig. 1.3 Importance of risk assessment in power generation

1.2.4 Importance of Risk Assessment in Climate Change

Climate change has become an important topic of discussion because of its alarming and potentially serious effects on planet Earth. The International Panel on Climate Change (IPCC) defines climate change as the change in the state of the climate that can be identified by the changes in the mean (and or the variability) of its properties which exist for a persistent period of time. One of the reasons for climate change is global warming. As the name signifies, global warming refers to the warming of temperatures around the world. This is caused by an increase in the greenhouse gases and other heat-absorbing gases such as carbon dioxide, methane, nitrous oxide etc., trapped in the Earth's atmosphere when it is heated by the sun [3]. From the figures that we have at hand, global warming is already having a serious effect on all the different factors of the environment on which humans depend in the course of undertaking their daily tasks. In order to prevent the effects of global warming from creeping into and disrupting the equilibrium for humans in their day-to-day life scenarios, adaptation is very important [4]. To facilitate this adaptation, steps need to be taken by which different types of *risks* are identified, assessed and managed, as shown in Figure 1.4. But it is important to note that the process of adaptation is a sequential and iterative one. The process of adaptation is achieved by first identifying the various events posing risks, assessing the level of their impact, and then developing policies by which they can be mitigated.

Fig. 1.4 Importance of risk assessment in climate change

1.2.5 Importance of Risk Assessment in Education Quality

The quality of education should be such that students will be able to meet challenges and cope with higher competitive levels when they are in the real world. The wave of globalization and advancement in technology and dynamic global changes has raised new public awareness which is becoming more powerful and inclusive in the 21st century. In assessing the quality of education educational institutions should ensure that the curriculum content is relevant to, and linked with, the requirements of industry and those of the public at large; and there must be cohesion between elementary, secondary and college/university education [5]. This relevance must be continuously updated and reviewed, in order to close any gaps that exist between the requirements of industry and the offerings of educational institutions. Educational institutions should also ensure excellence in teaching and learning processes, and that the teaching staff are highly trained and qualified in order to produce high quality graduates. Furthermore, in order to keep pace with the rapid changes in industry, educational institutions should adopt a competency-based curriculum. This kind of curriculum emphasizes performance, where students are encouraged to demonstrate their area(s) of competence, and their ability to perform a real job. Moreover, what is needed is a strict enforcement of these factors in the real world. Failure to incorporate these factors in day-to-day operations would put the educational institution at *risk* of not meeting the required educational quality, thereby not meeting the enrolment targets, not having financial sustainable benchmarks and negatively impacting its reputation, as shown in Figure 1.5.

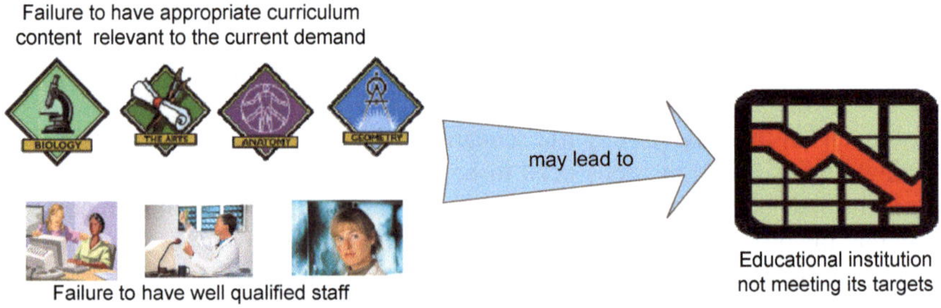

Fig. 1.5 Importance of risk assessment in maintaining educational quality

1.2.6 Importance of Risk Assessment in Software Project Development

A software project involves the development of new functionalities and successfully integrating them into the final desired product. However, achieving this in the real world is not easy as it sounds. Uncertainties and unexpected surprises could arise at various stages of the project which may lead to delays and failures. To address such uncertainties, effective project management is necessary at each step of the project development. Project management proactively assesses and ascertains the possibility of unexpected events occurring and takes steps by

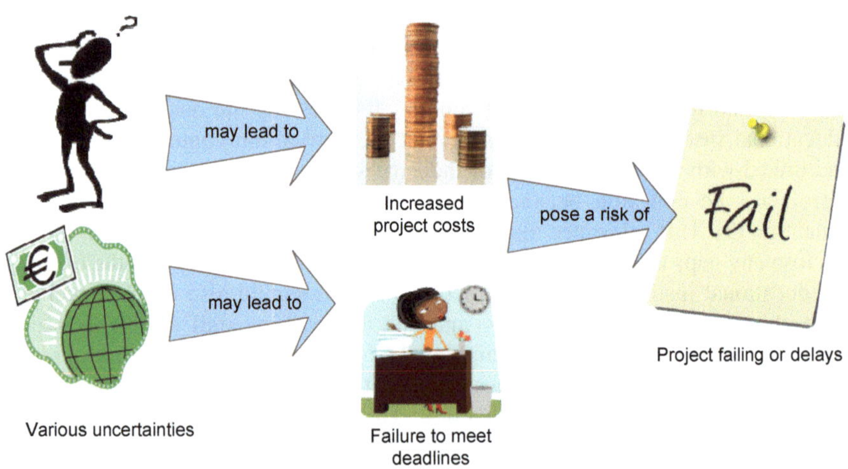

Fig. 1.6 Importance of risk assessment in software project development

which they can be avoided or mitigated to ensure the progress is as planned. In other words, project management effectively analyzes the *risk* of the project failing due to unwanted surprises and takes steps by which they can be avoided. As shown in Figure 1.6, failure to undertake this process at each step of project development would put at risk the timely completion of the project.

1.3 Understanding Risk

Even though each of the discussions above has its own domain meaning, one common thing that surfaces is the uncertainty associated with situations and events, from which risk arises. This inherent uncertainty arises from a *need* to achieve a certain outcome through an *activity*. From these discussions, risk can be seen as an outcome associated with a chain of events as follows: the need to achieve an outcome → which in turn gives rise to an activity → an activity may have different sub-activities → these sub-activities have uncertainty associated with them → the chance of an event occurring which may result in the occurrence of sub-activities being different from what is expected → possibility of risk in achieving the outcome. The links between these series of steps are shown in Figure 1.7.

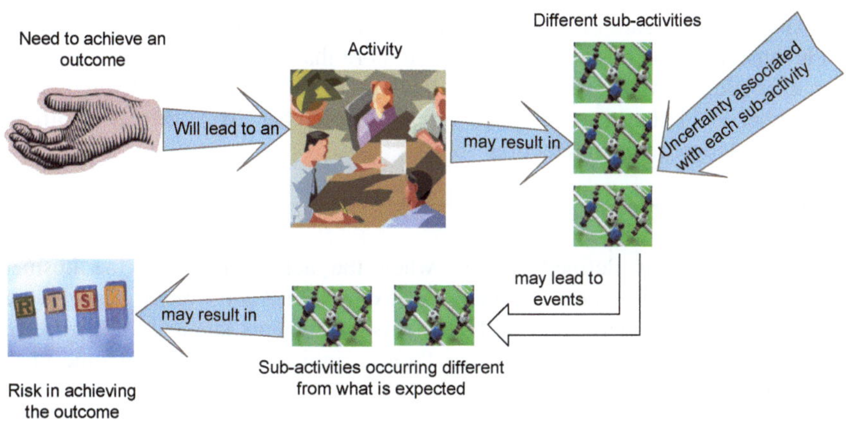

Fig. 1.7 Series of steps which many lead to risk

In most cases, it can be said that the event/s associated with risk will lead an interaction towards experiencing a loss in the activity, for example, the manufacturer not meeting the customers' demand due to production unit failure, the Genco not meeting the required energy demand due to downtime of generation units, an educational institution not meeting the required quality of education due to poor teaching staff etc. However, this does not mean that the presence of uncertainties or risk in an activity will always result in a negative outcome. There may be instances where the risk that arises from an uncertainty may result in an outcome that is beneficial, although initially it might have been leading the interaction towards a negative outcome. For example, the manufacturer may not meet the customer's demand initially due to failure of the production units, but accordingly, it develops strategies by which this is addressed and the customers' demand is met, more than expected. This can be due to the manufacturer using back-up goods or hedging the production with other third party manufacturers etc.

Such a change in the interaction's or an activity's outcome is achieved by analyzing the uncertainty associated with it and the actions or steps taken to avoid or mitigate them. But before an event in an activity can be classified as a risk, it is important to identify and analyze it according to the appropriate context of discussion and then classify it into different types of activity outcomes to which they can lead. In the next sub-section, we will explain this in more detail.

1.3.1 Possible Outcomes of an Interaction

As previously mentioned, each sub-activity or activity is responsible for having a collective final outcome in the business endeavor, which can broadly be divided into three categories as follows:

1.3.1.1 Positive Outcome

As shown in Figure 1.8, we define positive outcome as that when the achievements in the business endeavor are more than what was initially expected. For example, if Alice was promised that her order would be delivered on 23 September 2011, but was delivered on 15th September 2011, then it signifies a positive outcome in the interaction.

1.3.1.2 Neutral Outcome

Neutral outcome is defined as that when the achievements in the business endeavor are just what were expected. For example if Alice's order was delivered as promised on 23rd September 2011 but was a few hours late and this turned out not to have any serious consequences then it signifies a neutral outcome in the interaction.

1.3.1.3 Negative Outcome

Negative outcome is defined as when achievements in the business endeavor are below what was expected. This happens when, for example, Alice's order is delivered on 31st September 2011 as opposed to 23rd September 2011 – the day which was promised during the order confirmation.

To avoid the chance of a negative outcome occurring in the business activity, steps need to be taken to analyze and then avoid or mitigate the uncertainty and the events which lead to that outcome. However, before this is done, a pre-requisite is to identify the uncertainties and the event/s from an activity that have the *potential* to be classified as risk. This is further discussed in the next section.

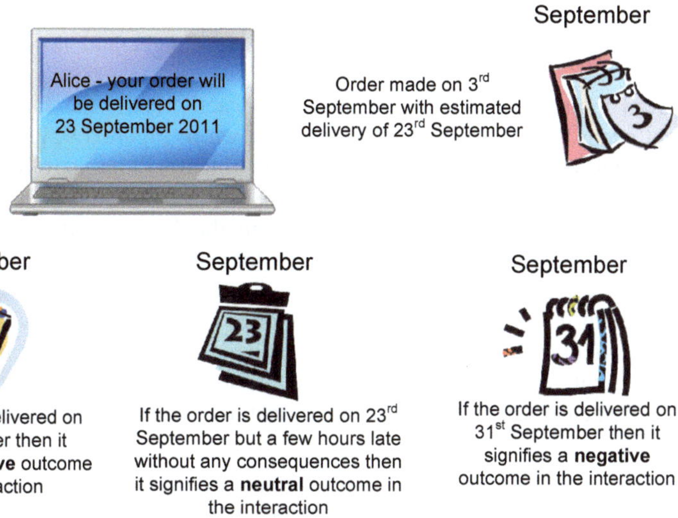

Fig. 1.8 Different types of outcome in the business activity

1.3.2 Classifying Event/s Which Arise from Uncertainty

As mentioned earlier, there are many different uncertainties that lead to different events that are possible in a business endeavor. But only those event/s which *might* have the capability to take the business endeavor in an unwanted direction or towards the occurrence of a negative outcome will be identified as risks. Note the association of the word 'might' in the above sentence. This is to take into consideration only those uncertainties and event/s that may have a chance of occurring, which in turn, will result in a negative outcome. By this, we are drawing a fine distinction between event/s that will *definitely* occur and those that *might* occur in a business interaction. This will lead us to classify the events which arise from uncertainty in two different categories: certain events and events posing as risk.

1.3.2.1 Certain Events

As previously mentioned, events which are classified as 'certain' will definitely occur in the interaction and will lead to a negative outcome. For example, let us consider the scenario discussed in Section 1.1. If a person drinks and drives through streets where police are checking for drink-driving, then he will *definitely* be caught and lose his driving license.

1.3.2.2 Events Posing as Risk

Events classified as 'associated with risk' may occur and might have the capability to take the business endeavor in an unwanted direction or towards the occurrence of a negative outcome. For example, let us consider a slight variant of the example above. If a person drinks and drives through the suburbs where the police *may* check for drink driving, then he risks being caught and may lose his driving license.

Even though the outcome of both these activities is the same, the way they might occur differs. In the first case, as shown in Figure 1.9, it is certain that the person will be caught as he knows that the police will be patrolling these specific streets; whereas in the second case, there is a chance of this event happening. Risk is used to categorize and define those event/s that have uncertainty associated with them. The uncertainty in this case is that the user wants the activity to occur in a particular way, but he is not sure of how or whether it will occur in that way. For example, in the second case, the person does not want the police to be present on those particular suburban roads on which he is driving, but there is uncertainty as to whether it is going to happen that way or not. This is different from the scenario in the first case where there is no doubt that the police are not going to be there.

Fig. 1.9 Difference between certain and uncertain events

Let us consider a different activity in the business domain. A Just-in-Time (JIT)-based electronic goods manufacturer determines that there is going to be an increase in customers' orders over the next fortnight due to the approaching holiday season. In order for the manufacturer to ensure that it meets demand in the required timeframe, all its production units must work at full capacity. During this timeframe, if the manufacturer intends to carry out the yearly maintenance of its production units forcing them to be offline, then it is definitely not going to meet the customers' demand in the *required time period*. On the other hand, if there are any unforeseen outages in its production units, there is a risk of the manufacturer not meeting the customers' demand in the *required time period*. The uncertainty associated with the second type of event is that the manufacturer wants to meet the customers' demand, but may not be able to do this due to the unforeseen outages. This is different from the certain events of the first case where the manufacturer will definitely not achieve the demand as the production units are offline. Such certain events *have* the capability to take the outcome of the interaction in a negative way, as compared to the uncertain events that *may* have the capability to do so. But it is very important for both these types of events to be identified, considered and managed for the successful completion of the business activity. Depending on the type of event, the process of dealing and managing them varies and hence they have to be considered and dealt with separately.

It is important that the occurrence of risk is analyzed and addressed to alleviate or minimize the *impact* and to successfully complete the business endeavor. In the next section, we discuss the impact that risk may have on a business activity as the result of its occurrence and why it is important that it be addressed.

1.3.3 Impact of Risk

As discussed earlier, an activity is carried out to achieve certain outcomes. Risk represents uncertainty and the chance of experiencing a negative outcome in, for example, the manufacturer not meeting the customers' demand, the educational provider not meeting its goals etc. However, apart from representing the chance of negative outcome, there is another effect which is a result of risk, that is, the '*impact*' it will cause in the interaction. Impact represents the effect that will be experienced as a result of not achieving the desired outcomes, for which the activity was being carried out.

If we consider the real world business endeavors discussed in section 1.2, in Figure 1.10 we can see that risk, apart from only representing the occurrence of a negative outcome also represents the impact as a result of that. For example, the risk of the manufacturer not meeting the customers' orders will result in it losing customers, market share to competitors etc. Similarly, the risk to the Genco in not meeting the electricity load may result in financial loss due to the non-commitment of bilateral contracts. This is similar in other real-world scenarios.

For example, let us consider the activity of aircraft maintenance and the possibility of a shortage in the required service parts to repair the aircraft. Risk in such an activity not only represents the chances of not getting the aircraft fixed due to a shortage in the required service parts, but also represents the impact of having the aircraft inoperable, for example, financial loss to the airline company for having to keep the aircraft on the ground, compensation for the affected passengers etc. Furthermore, the impact varies according to the context of the activity in which risk is being discussed. For example, in the scenario of aircraft maintenance, the risk of the aircraft being repaired incorrectly is that the aircraft may crash, resulting in loss of human lives. As shown in Figure 1.11, the impact in this scenario (loss of human life) is different from the financial loss to the

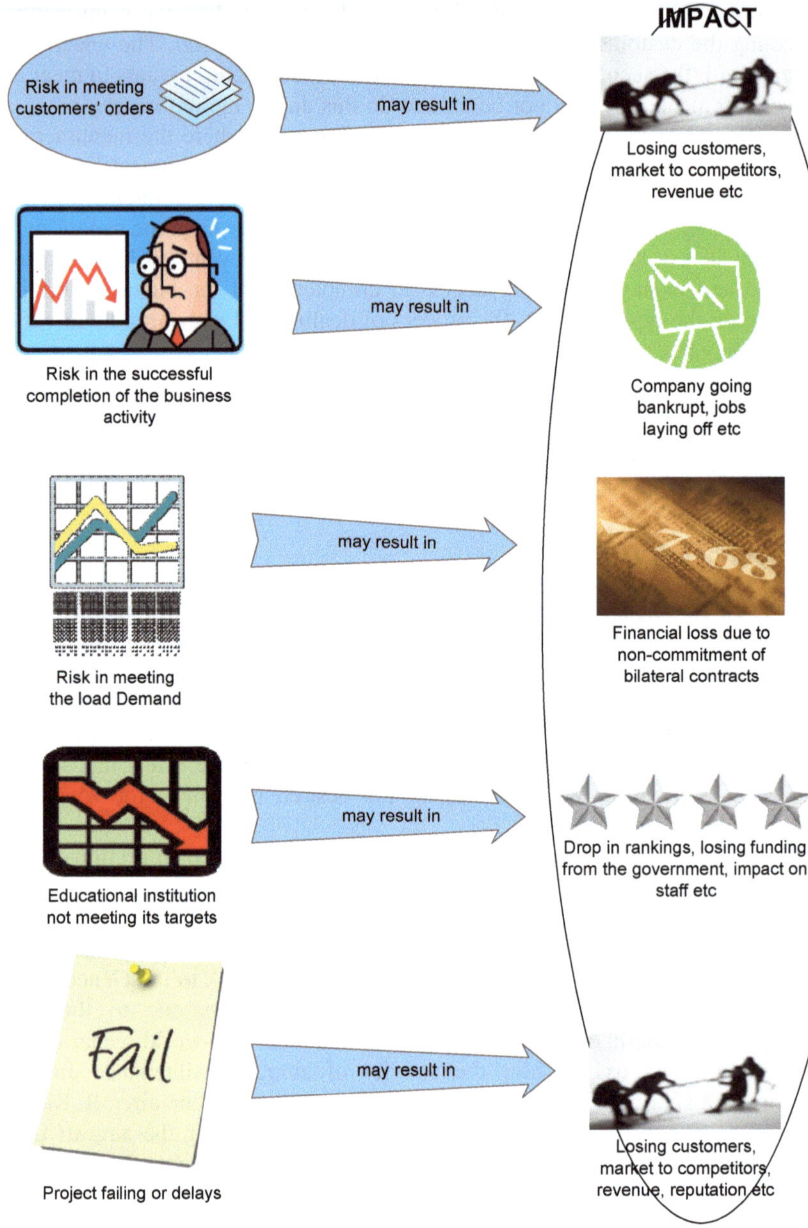

Fig. 1.10 Impact as a result of risk

airline company due to the aircraft being inoperable due to a shortage in the required parts. So with a change in the context of discussion, the impact of risk changes even though it still is applicable to the same activity, namely aircraft maintenance. This is discussed further in Chapter 6.

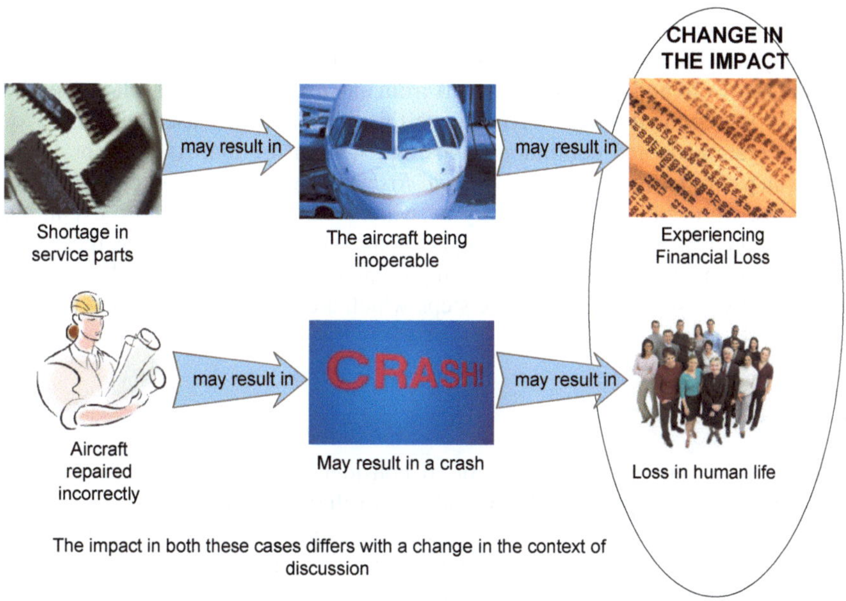

Fig. 1.11 Difference in impact with a change in the context of discussion

When risk is analyzed, it is important to ascertain the impact apart from just determining the chances of the uncertain events occurring, as this plays an important part in decision making in business endeavors. A survey of the literature revealed that there are numerous definitions of risk in different areas such as psychology [6], construction industry [7], business etc. As these different areas look at different domains of application, it is not surprising to note that the definitions of risk too vary across these applications. But no matter how risk is defined, all these definitions will directly or indirectly highlight the occurrence of uncertainty and the associated impacts from them. But the way they are expressed and defined varies according to how they are best understood in the specific context of the domain. So, an important point to note here is when analyzing risk, it is important to first identify how risk is defined and understood in that domain before analyzing it. In the next section, we discuss the factors which influence Risk.

1.4 Important Factors Which Influence Risk

By now you would have understood that given the unwanted outcomes that it brings along with it, it is best if risk does not occur in the interaction. But, in order for this to be a reality, an interacting user cannot just sit and hope that risk does not occur in their business activity. In order to ensure that this is the case, the interacting user has to be on their guard and hence act accordingly. So what do we mean by *'on their guard'?* This is very much like an interacting user taking a shield and sword and protecting himself from any event that will result in risk. But it is not as easy as it sounds. To achieve this, an interacting user must carry out various systematic and continuous steps that will ensure that risk is kept at bay. These steps are termed *Risk Analysis and Risk Management.* But to carry out such a process, we need a systematic way by which we first identify those events that are associated with risk and later analyze them. We will further discuss these steps in the next chapter. There are some important factors associated with risk to be considered whilst undertaking these steps which are discussed in the remainder of section 1.4.

1.4.1 Uncertainty

Uncertainty is an important concept to consider when defining risk. In broad terms, uncertainty represents the probability of the interacting user not arriving at the definite outcome that they want to achieve as a result of an event, which is a part of his business activity. As shown in Figure 1.12, uncertainty represents an interacting user who wants to achieve a certain desired outcome in a business activity but is uncertain and unsure about whether it will actually be achieved or not. This uncertainty may lead towards experiencing risk in an interaction and hence is an important concept to consider when identifying risk in a business activity. If the interacting user is sure that he will definitely achieve his desired outcome from an event, then there is no uncertainty associated with it and hence no risk. On the other hand, if the interacting user is sure that he will definitely not achieve his desired outcome from an event, then there is no uncertainty associated with it and hence it is not termed as risk but as certainty. So only those events that have uncertainty associated with them are classified as risk and hence uncertainty is an important concept to consider when identifying risk.

1.4.2 Possible Outcome

The outcome of an event is also important in classifying that event as being risky or not. As discussed earlier and as shown in Figure 1.13, there can be three possible outcomes at the completion of a business activity; namely positive, negative and neutral. The types of outcome that are associated with risk are the occurrence of those uncertain events that will lead to a negative outcome or those that will lead to the non-occurrence of a neutral or positive outcome in the business activity. So the interacting user, when identifying risk, should identify

Fig. 1.12 Uncertainty is an important factor of risk

those uncertain events that will lead to such negative outcomes in the business activity. It may be possible that those uncertain events which the interacting user identifies as associated with risk in a business activity will eventually not lead to the occurrence of a negative outcome, or will even lead to the occurrence of a positive outcome. To carry out the process of risk analysis, it is imperative that the interacting user should consider all those events that he initially identifies as having the capability of driving the outcome of the interaction in the negative direction, irrespective of what type of outcome they eventually have at the end.

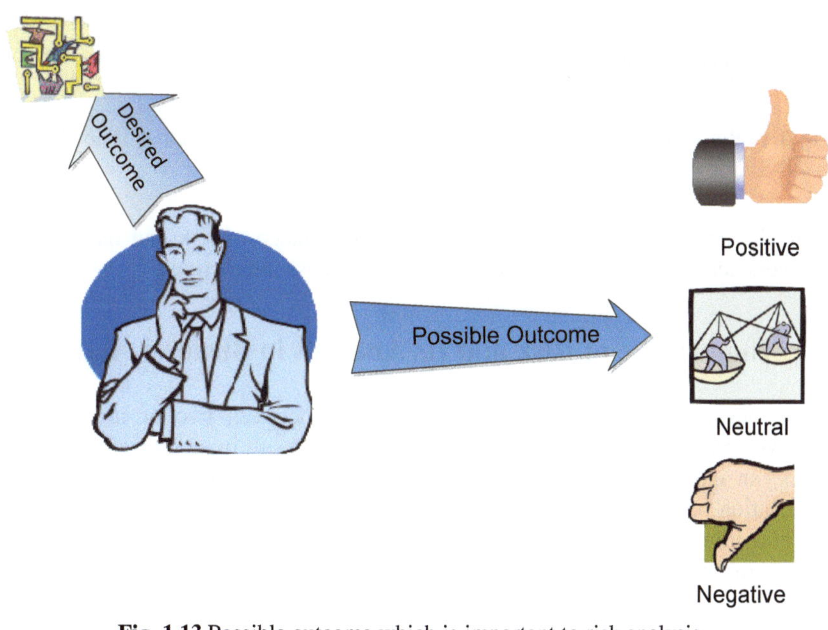

Fig. 1.13 Possible outcome which is important to risk analysis

1.4.3 Time

Time is another important characteristic to consider when identifying risk in a business activity. This is because risk is dynamic and it is possible for an interacting user to have different impressions of another user in the specific factors, at different points in time. For example as shown in Figure 1.14, interacting user 'A's past business association with company 'AB' had a positive outcome. But by considering company 'AB's recent interaction history over a given period of time, the interacting user 'A' might now consider the events as having uncertainty associated with them which has the capability of resulting in a negative outcome in a business activity. So, while carrying out the process of risk analysis, is it important for the interacting user to consider the notion of time and identify the different events that will have uncertainty associated with them, at recent different points in time.

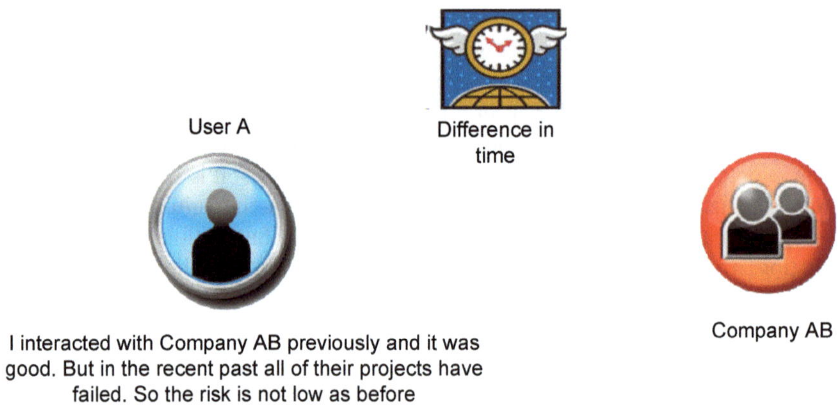

Fig. 1.14 But in the recent past, all of their projects have failed, so the risk is not AS low as before

In the next section, we will discuss the importance of risk and its analysis in business activities with a special focus on e-business interactions.

1.5 Importance of Risk and Its Analysis in Business Activities

Some business activities involve collaboration between two or more users to achieve certain specific business outcomes, which result in an *interaction.* In most cases, an interaction is both an association and collaboration. Each user in the association has certain objectives to achieve, which is the main motivating factor for him joining the collaboration. But to achieve these outcomes and to have a successful collaboration, it is imperative that all users behave or commit to behaving in a way that is expected. The failure of any user to commit to what they should actually do or what is expected that they will do, will likely lead to the failure of the interaction and the subsequent occurrence of a negative outcome. Such categories of outcome/s may be undesirable to one or more of the interacting users and subsequently they would want to avoid these.

In the modern world, e-Business has rapidly emerged as an important architecture for facilitating business activities. It has provided users with sophisticated technologies which ease the process of carrying out their activities. This enables them to complete their tasks in less time and reduce the delays associated with conventional methods of business interaction, while at the same time, boosting their efficiency and helping them improve their sales and productivity. An important characteristic of e-business interactions is that an association or collaboration is formed between any two users who might not be physically in front of each other, but may possibly be miles apart. In this regard, this type of interaction is quite different from traditional business interaction. As a result, in such types of interactions, it may be more difficult to make a judgment about the users' existence and their credibility as compared to the physical users that are available in a direct physical interaction. In addition, one is unable to access physical body language and the facial cues of other users. Business-to-Business (B2B), Business-to-Consumer (B2C) and Consumer-to-Consumer (C2C) are the architectures of this type that are widely utilized to facilitate e-business interactions. But, in spite of such characteristics, the demand for such type of interactions and e-commerce business in particular is growing because of its robust nature [8]. According to a report from the Australian Bureau of Statistics updated in 2006, the number of Australian consumers who engaged in online shopping in the year 2000 grew by a mammoth 66 % compared with the figures from the previous year when 1.3 million Australians ordered or purchased goods over the Internet during the year [9]. The increasing demand for internet shopping is not only limited to the Internet users in Australia. According to the Dutch home shopping group, in the year 2006, Dutch nationals bought goods worth 4 billion dollars (2.92 billion Euros) via the Internet, an increase of 28 % from the figures of the previous year [10]. A report by Forrester research showed that in 2003, about 40 % of online European consumers actually purchased online, up from 19% in 2000, and in 2006, online shopping in Europe was on the increase, especially in countries like Sweden and Netherlands [11]. In the United States, business retailers encourage their customers to shop online for their products and are bringing the Internet into their stores in the form of web kiosks that allow their customers to shop online at the store's e-tail counterpart [12].

Apart from the advantages provided by such architectures which result in their fast adoption, there are some very important considerations for users, such as notions of 'security', 'trust' and 'risk', which help to make the interactions over this paradigm safe, secure and informed. This is supported by Chang et al. [13] who state that 'the dynamic, open and convenient web environment not only boosts business potential and the economy but also creates concerns of security, trust, privacy and risks'. So, before utilizing the facilities provided for their advantage, users should consider and analyze these concepts which will lead them making an *informed interaction-based decision*. In the next sections, we define what is meant by an informed interaction-based decision and will discuss each of the above mentioned concepts, the assessment of which will help an interacting user make an informed decision as to whether to join or form a collaboration between users. As our focus in this book is risk, we will discuss the importance of analyzing and assessing risk in order to maximize the expected output in a business endeavor.

1.6 Informed Interaction-Based Decision

As discussed in the last section, the Internet has been widely accepted by users as a global communication infrastructure [14]. Due to the ease of use and dependence of the users on the provided functionalities, some users become negligent and use the facilities without weighing the benefits with the associated outcomes. It is common in today's world to find deceitful users who try to exploit others for their own benefit. The fact that such e-business interactions are conducted in a widely-distributed, open and decentralized environment makes the tasks of such deceitful users much easier. Some careless users pay the price for their negligence but some users are skeptical and want to weigh the advantages and disadvantages before deciding on an interaction, and hence need some framework which can help them to achieve this and make an *informed interaction-based decision*. In the context of the discussion of this book, an informed interaction-based decision is defined as a proactive decision that will lead to the achievement of the goals for which the collaboration in the business endeavor was formed. By such informed decision making, the interacting user can reduce the potential of forming a business collaboration with deceitful users and carefully choose only users who are unlikely to be deceitful to form an interaction with, hence maximizing their interaction experience and benefits. This is shown in Figure 1.15.

Let us consider the scenario of a user buying a PlayStation 3 from an e-marketplace. There is probably more than one seller who is willing to sell the PlayStation 3 to the buyer, and from these sellers, the buyer has to choose one. Failure of the buyer to achieve any of his desired outcomes will mean that he will experience a loss in his business endeavor that is expressed by a negative outcome. So, in order to achieve the desired outcomes, the buyer has to make an *informed decision* to choose the best possible seller by whom the best possible outcome in the interaction can be achieved. This will be done by analyzing the various concepts that will help the buyer to believe that he can achieve what he wants in interacting with that seller. These concepts are security, trust and risk. In the next sub-sections, we will discuss the role of 'security', 'trust' and 'risk' in making an informed interaction-based decision in the e-business domain, the issues they address and their impact in enabling the business endeavor to be carried out in a safe, secure and informed manner.

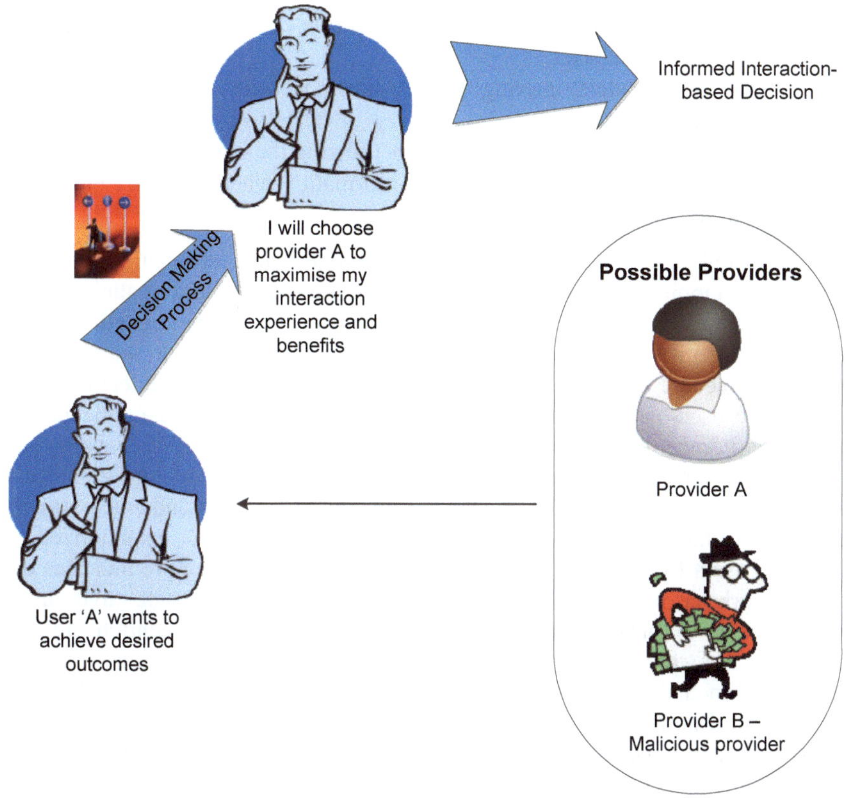

Fig. 1.15 Informed interaction-based decision

1.6.1 Role of Security in Making an Informed Interaction-Based Decision

Security is a crucial aspect in any interaction or collaboration. It is the act of taking appropriate measures by which the threats or the dangers to the interaction from unauthorized users or environments are addressed, checked or eliminated. In other words, security means securing or protecting the interaction medium and the users of the interaction from anything which might present a threat to the individuals involved in it. In today's technologically advanced world, the Internet has enabled information to be exchanged either by the user or the business in a much more efficient, fast and reliable way. This information can be sensitive and forms a very important resource for both businesses and users. So, in addition to keeping such information safe, as interactions are conducted over the virtual medium, users also have to utilize security mechanisms during the exchange of information with other users. There are various ways by which such information may be vulnerable to attacks due to a lapse in security. Some of these, to mention a few are:

- Hackers or intruders gaining access to sensitive customer information stored in the business databases;
- Hackers or intruders altering the business website and providing false information to the users;
- Credential management among the users utilizing the business resources;
- Hackers eavesdropping on the users' information when the latter enter their financial information; etc.

Various techniques such as eavesdropping, masquerading information, denial of service, key tapping, the use of trojans or viruses are available to malicious users to achieve this. Avoiding such unwanted activity is achieved by strengthening the security mechanisms. In the context of this book, security relates to the process of providing a secure environment or a secure space in the open architecture of the web, where users can exchange information with others within the collaboration with an assurance that the information which they share and utilize is being sent only to and by the intended users, and that the information is not being changed in any way. As shown in Figure 1.16, this is achieved by utilizing and employing the security mechanisms that will prevent hackers, attackers, unauthorized or malicious users from carrying out any such activities. The literature indicates that a lot of work has been done in the area of security to provide a safe environment and to act as a barrier to malicious users. Some worthy of mention are data encryption or providing a secure safe environment by using improved authentication and validation techniques [15, 16], digital signatures [17], and access control policies [18, 19] etc. All these techniques will help the users of the collaboration to concentrate on the main aspects related to their interaction, by providing them with the reassurance that the issues related to security are being looked after by the security mechanisms. But, as the Internet is ever developing and fast changing, the various techniques by which the hackers or the un-authorized users can impact on secure communication between any two collaborating users is also ever evolving. Hence, the presence and the development of security technologies for combating such techniques by hackers are essential and should be continuous.

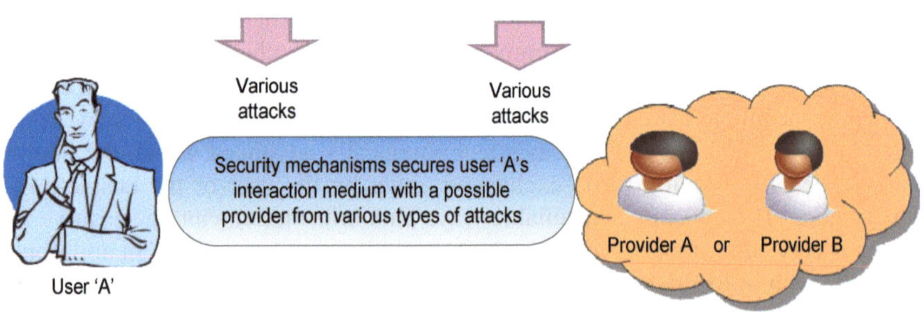

Fig. 1.16 Role of security in making an Informed Interaction-based Decision

So, security during online business is utilized to efficiently shield the interaction environment so that users can collaborate without disruptive forces affecting their interaction. In other words, security is the first aspect considered by users who take part in an online interaction as it acts as a barrier to the outside forces which aim to disrupt the interaction; and enables the users of the interaction to initiate the process of ascertaining whether or not to place his/her 'trust' in the other interacting user/s.

1.6.2 Role of Trust in Making an Informed Interaction-Based Decision

The word 'trust' in the previous discussion can be interpreted in different ways according to its context. The concept of 'Trusted Computing' has been proposed in the literature in an attempt to solve some of the security problems encountered in today's world by adopting security mechanisms [20]. Trusted Computing is a part of the trusted computing group which is formed by an alliance of Microsoft, Intel, IBM and HP in providing a platform which cannot be tampered with, and where the applications can communicate securely with each other [21]. Considering this in the context of the business domain, the word 'trust' is related to trusting a business or interaction infrastructure based on the security mechanisms that it provides, or trusting the security mechanisms of a collaboration. It refers to the perception of a user in 'trusting' or, in other terms, 'considering' the other user/s as a possible entity with whom they may decide later to interact or not. This notion of trust is quite different from that which is required in the context of a business endeavor when a user actually decides to interact with the other user/s, and where we mean that 'a user trusts another user in the following interaction'. In this context, the word 'trust' refers to the level of confidence that a user has in the other user's ability and willingness to achieve his desired outcomes through the interaction.

Although the word 'trust' is used in both instances, it assumes quite a different meaning in each instance, due to the variation of the context of discussion. In the context of trusted computing, trust refers to the user accepting the provided security mechanisms adopted in the interaction infrastructure by which he feels safe and not vulnerable to the outside forces which might hamper its interaction. But this level of trust does not in any way mean that the user will interact with the other users of the collaboration that satisfies the level of required trust, or in other words, this level of trust in no way provides a guarantee or assurance that the other user will act in a way by which the desired outcomes of the interaction will be achieved. In order for a user to have such level of trust necessary to really commit himself to an interaction with other users, he needs to believe that by the collaboration, he can achieve his desired outcomes. In this context, the notion of trust is different from the security provided by the interaction infrastructure as it refers to analyzing the other users' reputation, based on their previous actions and as a result, determine his level of confidence or belief in their capability to provide what the interacting user wants as a result of the interaction. Such a level of trust (as shown in Figure 1.17) is one of the important factors required to make an informed interaction-based decision.

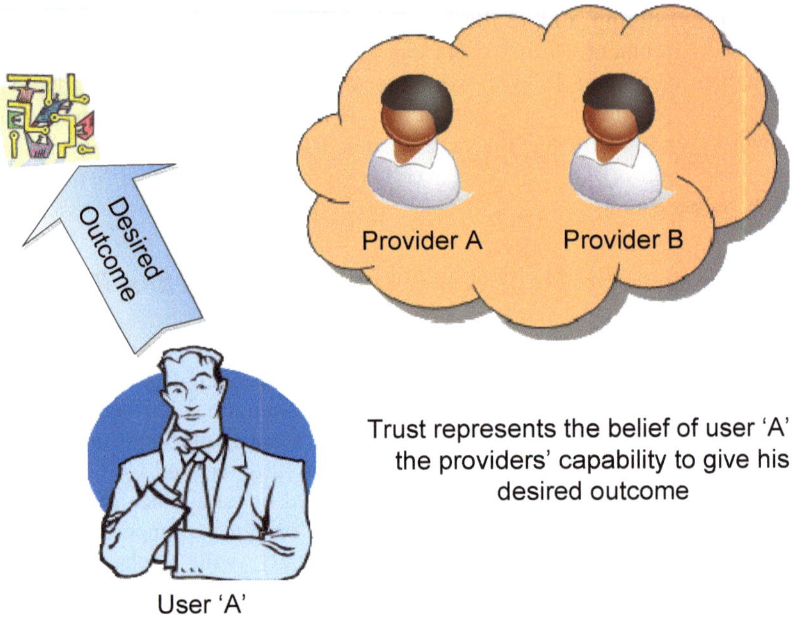

Fig. 1.17 Role of trust in making an informed interaction-based decision

From the above discussion, it can be concluded that a user takes the initial step of ascertaining the level of security provided by the interaction infrastructure before deciding whether or not to enter into collaboration, as security helps to provide a safe environment within which an interaction can be confidently carried out. Based on the level of security, the user can determine his level of belief to make an informed decision about his interaction with the other users.

1.6.3 Role of Risk in Making an Informed Interaction-Based Decision

Risk during business decision-making highlights the threats and the associated impacts in the interaction. These threats are not only limited to the ones that are addressed by the analysis of security but may also cover those which arise from deceitful users in collaboration. So, the analysis of risk is important in business collaboration.

As there is an inter-relationship between security and trust as discussed earlier, similarly there is an inter-relationship between risk-security and risk-trust. As shown in Figure 1.18, even though the analysis of risk gives different and important representations compared to trust and security, there is confusion in the literature that considers either trust [22] or security [23] to be the authoritative concept on risk, and that based only on the level of security and trust can the decision-making process in the business collaboration be carried out; as it also represents or nullifies the level of risk in the interaction. But, in reality, this is not the case. Security, trust and risk in business activities represent different concepts, depending on the different areas of the business endeavor in which they are targeted. In generic terms, security in the context of business collaboration represents the mechanisms to safeguard the interaction medium; trust shows the level of belief that a user has in the capability and desire of the other user to provide him with his desired outcomes. But such secure medium and the level of belief does not express the uncertainties, degree of threat and the magnitude of loss that could possibly be experienced in the interaction. Such analysis can be determined only by the assessment of risk in the interaction in that context. So, the individual analysis of security, trust and risk is very important in order to make an informed interaction-based decision in e-business collaboration. We discuss about the different types of risks in the next section.

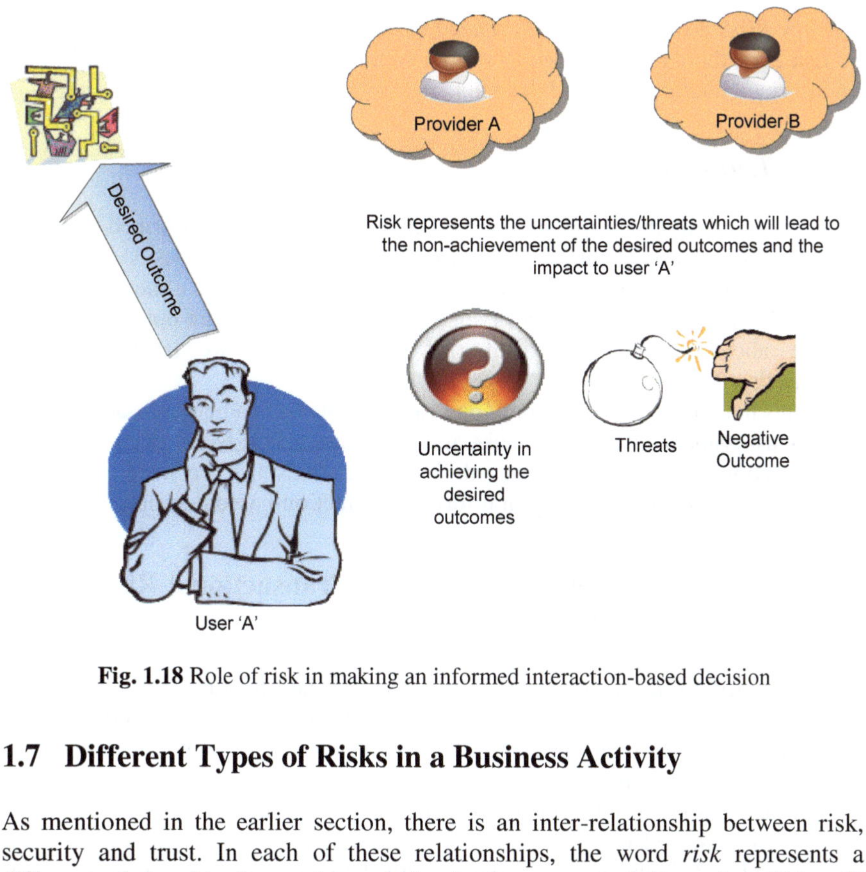

Fig. 1.18 Role of risk in making an informed interaction-based decision

1.7 Different Types of Risks in a Business Activity

As mentioned in the earlier section, there is an inter-relationship between risk, security and trust. In each of these relationships, the word *risk* represents a different relationship due to the variation in the context of discussion. This will lead to a definition of the different types of risk that could be present in a business interaction as explained in the next sub-section.

1.7.1 Security Risk

In the context of a business interaction, analyzing risk in the area of security will only ascertain the *Security Risks* present in the business endeavor. Such risks, even though important, only highlight the security threats of the collaboration and do not provide the required complete picture to make an informed interaction-based decision. In such scenarios, even if a user considers the level of security risk as acceptable, based on that level of risk, he cannot make an informed decision about forming a collaboration with another user, as the type of risks which may be encountered in doing so are much different from only security risks.

1.7.2 Privacy Risk

Privacy Risks are privacy threats related to the business collaboration. Such types of risk even though are important, do not capture all the threats required to make an informed interaction-based decision.

1.7.3 Transactional Risk

Awareness of another type of risk which is important for informed decision-making is in relation to threats (apart from security and privacy risks) which lead to the non-achievement of business outcomes. This may be either due to the deceitful nature of the other user or some other events which are outside the scope of the abovementioned types of risk. In this book, we term such risks *Transactional Risk.*

In this book our focus is on only one category of risk, namely transactional risk and its assessment in order to make informed interaction-based decision. In the next section, we examine the different relationships between trust and transactional risk while making an informed interaction-based decision.

1.8 Relationships between Trust and Transactional Risk in a Business Activity

As mentioned earlier, it has been often argued in the literature that transactional risk is an important factor in deterring a user from shopping online on the Internet [14, 24-26]. To overcome this, researchers have tried to combat the effect of transactional risk in the interaction by the notion of trust. Various conceptualizations of the term 'trust' while making an interaction-based decision have been presented in the literature. A generic definition of trust adopted from Chang et al. [13] in the area of business is 'the belief or faith that a person or agent has in the another person or agent with respect to certain activities at a given time'. But apart from this, another important perception to consider in such interactions is the notion of transactional risk which identifies the level of threat, failure and the possible losses in the interaction.

In order to explain the collective importance of both trust and transactional risk in a business interaction, consider a practical example from our daily lives. Before collaborating with a user and investing our resources in interacting with them over a virtual environment, it is reasonable for us to consider questions such as:

- To what extent can I rely on the other user/s?
- Will the other user/s be able to deliver what I want?
- Will I lose the resources that I am investing?
- To what degree will the other user/s not deliver what I want?
- What will be the degree of loss for me in the interaction?

Based on our perception of the answers to these questions, we decide our future course of action in the interaction. In other words, the questions that we ask ourselves before initiating the interaction are related to the trust and the transactional risk associated with the interaction, and its analysis impacts upon our willingness to interact, or on our decision regarding our future course in the interaction. Trust and transactional risk are two concepts associated with an interaction, which complement each other when making an informed decision about the future course of action. Both of these concepts, although complementing each other, express different meanings, which in turn are not interchangeable and cannot be substituted in achieving their respective outputs. So each of them must be individually analyzed and considered when making an informed decision about an interaction. Hence, it is incorrect to compare them, or decide which one of them is the more important for better decision making in an interaction, or try to alleviate the effect of one concept by the other in a business interaction. In the literature, Mayer et al. [27] highlight the confusion in the relationship between risk and trust by stating "it is unclear whether risk is an antecedent to trust, is trust, or is an outcome of trust". Trust and risk can have different types of effect on the user depending on the relationship between them. As summarized by Gefen et al. [28], the relationship between trust and risk in an interaction can be classified into three main broad categories, as discussed in the following sub-sections.

1.8.1 Moderating Relationship

In such relationship, transactional risk moderates the effect that the user's trust has on his behavior or on his willingness to buy or to proceed with the interaction. In other words, this relationship mentions that trust is relevant when transactional risk is high and vice versa.

1.8.2 Mediating Relationship

In this type of relationship, trust and transactional risk both have a mediating effect on the behavior of the user in the interaction. In other words, the level of trust scales the effect of transactional risk on the behavior of the user in the interaction. This is supported by Jarvenpaa and Tractinsky [29] who state that trust affects the transactional risk which in turn, affects behavior.

1.8.3 Independent Relationship

This type of relationship represents an independent effect of both trust and transactional risk on the behavior of the user in the business activity. Kim and Prabhakar [30] consider this type of relationship when studying the behavior of the user in the context of internet banking.

So, no matter what relationship is considered by the user between trust and transactional risk, from the above discussion it can be concluded that both these concepts are important and play a significant role in deciding his behavior in forming the collaboration, even though their conceptualization may be different. In the next section we discuss the difference in the outputs of trust and transactional risk.

1.9 Difference in the Outputs of Trust and Transactional Risk Analysis

As their conceptualization, the way trust and risk are quantified and the outputs that they give also vary. Trust in an interaction is mostly expressed in probabilistic terms that represent the level of belief, whereas transactional risk is best understood when expressed in terms of probabilities under conditions of uncertainty that represents the failure of an interaction and any possible loss. As both these concepts are expressed in terms of probability, it may be possible that they both express the probability of an outcome related to a certain event, but the representation of the outcome and types of outcomes that each of them represent are different. What trust analysis will not give are the negative aspects related to the interaction and the impact associated with it. This is given by the analysis of transactional risk. In this regard, the terms ‘trust’ and ‘transactional risk’ are concepts whose evaluation would give two contrasting perspectives of occurrence of a same event. But in broad terms, the manner in which they are analyzed, and their outcomes, are not simply the converse of each other. Rather, they provide various insights which might help to resolve or allay the initiating user’s various doubts or suspicions that are present prior to making an informed decision.

With the evolving and ever developing technology of the modern age Internet for facilitating e-commerce interactions, users would want to analyze the detailed aspects of their interaction before making an interaction-based decision. They would rather avoid deciding on an interaction from a one-dimensional perspective analysis of the interaction, i.e. only based on the level of belief or trust that a user has in interacting with the other user; rather, they would surely also want to take into consideration each possible outcome and the associated costs to them, while making an informed interaction-based decision. This is done by analyzing the level of failure, along with the costs associated with those failures, in the interaction, apart from just considering the level of belief in the interaction. Measuring the loss or impact in the interaction is within the scope of transactional risk in the interaction, which cannot be compensated for by analyzing the degree and level of trust. Hence, analyzing transactional risk in the interaction is very important when making an informed interaction-based decision, apart from considering only the level of trust.

1.10 Recapitulation

This chapter provides the background for the following chapters. In this chapter we first discussed risk, its application in various different real world applications, its characteristics and the factors on which it depends. We then focused on the importance of risk in business interactions and studied the concepts of security, trust and risk that are essential and play a pivotal role in helping the interacting user to make an informed interaction-based decision. We discussed in detail each concept and highlighted the difference between their analysis to stress the importance of risk and its analysis while decision making in business interactions. We also discussed the different types of risk in a business interaction and then highlighted the specific areas which should be analyzed when making an informed interaction-based decision in business activities. In the next chapter, we discuss the process of transactional risk analysis in business activities.

References

1. Hussain, O.K., Dillon, T., Hussain, F., Chang, E.: Probabilistic Assessment of Loss in Revenue Generation in Demand-Driven Production. Journal of Intelligent Manufacturing (2011), doi:10.1007/s10845-011-0518-4
2. Mateus, J.C., Cuervo, P., Jannuzzi, D.P.: Hydrothermal bilateral contract negotiation with energy and financial risks analysis. In: IEEE Power and Energy Society General Meeting - Conversion and Delivery of Electrical Energy in the 21st Century, pp. 1–8. IEEE, Pittsburgh (2008)
3. `http://www.scidev.net/en/features/fossil-fuels-the-evolution-of-human-culture.html`
4. Perdomo, A., Hussain, O.: A multisystem climate change adaptation approach for water sustainability in regional Australia. International Journal of Global Warming 3, 39–54 (2011)
5. Smith, D.B., Fletcher, H.D.: The Balance Scorecard: A Useful Framework for Achieving Curriculum Integration. Journal of Executive Education, 41–55 (2004)
6. Spink, M.J.P., Menegon, V.M., de Souza Bernardes, J., Coêlho, A.E.L.: The Language of Risk in Psychology: A Social Constructionist Analysis of a Psychological Database. Interamerican Journal of Psychology 41, 151–160 (2007)
7. International Marine Contractors Association: Identifying and Assessing Risk in Construction Contracts (2006)

8. Oh, W.: C2C Versus B2C: A Comparison of the Winner's Curse in Two Types of Electronic Auctions. International Journal of Electronic Commerce 6, 115–138 (2002)
9. Australian Bureau of Statistics: Online Shopping Blooms. Use of the Internet by Householders, Australia (November 2000)
10. Thuiswinkel Organisatie: Internet shopping gaining popularity in the Netherlands (2007), http://lifestyle.monstersandcritics.com/life/news/article_1334910.php/Internet_shopping_gaining_popularity_in_the_Netherlands
11. Reitsma, R.: European Online Shopping Landscape 2007. Forrester Research (2007)
12. Enos, L.: E-tail Invades the Real World. E-Commerce Times (2001), http://www.ecommercetimes.com/story/7406.html
13. Chang, E., Dillon, T., Hussain, F.K.: Trust and Reputation for Service-Oriented Environments. John Wiley & Sons, Ltd., West Sussex (2006)
14. Su, B.-C.: Risk behavior of Internet shopping: Comparison of College Students' versus Non-Student Adults'. In: Proceedings of the 5th International Conference on Electronic Commerce (ICEC 2003), pp. 181–185. ACM, Pittsburgh (2003)
15. Tian, B., Han, S., Hu, J., Dillon, T.: A mutual-healing key distribution scheme in wireless sensor networks. Journal of Network and Computer Applications 34, 80–88 (2011)
16. Gao, L., Chang, E., Parvin, S., Han, S., Dillon, T.: A Secure Key Management Model for Wireless Mesh Networks. In: 24th IEEE International Conference on Advanced Information Networking and Applications, pp. 655–660. IEEE, Perth (2010)
17. Needham, R., Schroeder, M.: Using Encryption for Authentication in Large Networks of Computers. Communications of the ACM 21, 993–999 (1978)
18. Cvrček, D., Moody, K.: Combining Trust and Risk to Reduce the Cost of Attacks. In: Herrmann, P., Issarny, V., Shiu, S.C.K. (eds.) iTrust 2005. LNCS, vol. 3477, pp. 372–383. Springer, Heidelberg (2005)
19. Dimmock, N., Belokosztolszki, A., Eyers, D., Bacon, J., Moody, K.: Using trust and risk in role-based access control policies. In: Ninth ACM Symposium on Access Control Models and Technologies, New York, pp. 156–162 (2004)
20. Schoen, S.: Trusted Computing: Promise and Risk (2003)
21. Anderson, R.: 'Trusted Computing' Frequently Asked Questions (2003)
22. Aberer, K., Despotovic, Z.: Managing trust in a Peer-2-Peer Information System. In: Proceedings of the Tenth International Conference on Information and Knowledge Management, CIKM 2001, pp. 310–317, Atlanta, Georgia, USA. ACM (2001)
23. Datta, A., Hauswirth, M., Aberer, K.: Beyond "web of trust": Enabling P2P E-commerce. In: IEEE International Conference on E-Commerce, CEC 2003, pp. 303–312. IEEE Computer Society Press, Newport Beach (2003)
24. Rao, H.R.: Study Shows Perceived Risk of Online Credit Purchases Linked to Trust, Familiarity with Intermediaries. University of Buffalo. News Centre (2007)
25. Grazioli, S., Jarvenpaa, S.L.: Perils of Internet Fraud: An Empirical Investigation of Deception and Trust with Experienced Internet Consumers. IEEE Transactions on Systems, Man and Cybernetics 30, 395–410 (2000)
26. Hoffman, A.M.: A Conceptualization of Trust in International Relations. European Journal of International Relations 8, 375–401 (2002)
27. Mayer, R.C., Davis, J.H., Schoorman, D.F.: An Integrative Model of Organizational Trust. The Academy of Management Review 20, 709–734 (1995)

28. Gefen, D., Srinivasan Rao, V., Tractinsky, N.: The Conceptualization of Trust, Risk and Their Relationship in Electronic Commerce: The Need for Clarifications. In: Proceedings of the 36th Hawaii International Conference on System Sciences, HICCS 2003, pp. 192–201. IEEE, Big Island (2003)
29. Jarvenpaa, S.L., Tractinsky, N., Saarinen, L.: Consumer trust in an Internet store: A Cross-Cultural Validation. Journal of Computer-Mediated Communication 5, 45–71 (1999)
30. Kim, K., Prabhakar, B.: Initial Trust, Perceived Risk, And The Adoption of Internet Banking. In: Proceedings of the Twenty First International Conference on Information Systems, Brisbane, Australia, pp. 537–543 (2000)

Chapter 2
Transactional Risk Analysis in Business Activities

2.1 Introduction

A substantial body of literature, based on rational economics, argues that the decision to buy is based on risk-adjusted cost-benefit analysis [1]. So as mentioned in the last chapter, transactional risk is one of the important factors in business decision-making. Furthermore, the importance of transactional risk is not just limited to business interactions, but is evident in almost all activities conducted in the world today, regardless of the domain, including areas such as health, mining, safety, air travel, planning and infrastructure etc. But regardless of its importance, it is incorrect to think that the term *transactional risk* can be defined globally or in a generic way, which can then be applied to any discipline of discussion. As seen from the discussion in Section 1.2, even though the broad meaning across domains may be the same, the specific objects of analysis and output concepts represented by the term *transactional risk* vary across disciplines. This means that the definition of *transactional risk* in one discipline may not be suitable for other disciplines, as their object of analysis might vary with the change of discipline, thereby changing its interpretation and the sub-categories of its analysis. Sub-categories of analysis are those specific factors by which *transactional risk* is determined and expressed in that domain.

So, before analyzing transactional risk in a domain, it is important to understand the object of analysis in that domain and identify its sub-categories, accordingly. In this chapter, we identify the sub-categories and characteristics according to which transactional risk should be analyzed in business activities. We will then study the different types of uncertain events related to transactional risk and the process of risk analysis.

2.2 Subcategories for Transactional Risk Analysis in Business Activities

As mentioned in Chapter 1, transactional risk in business domains expresses two important aspects related to the activity. They are the probability of failure in achieving the desired outcomes and the impact or consequences of failure of the

O.K. Hussain et al.: Risk Assessment & Management in the Networked Economy, SCI 412, pp. 31–53.
springerlink.com

business activity. Both these aspects are important while transactional risk analysis as can be explained by considering the empirical evidence in a web-based sale activity. In such activities, for higher cost items, the web usually does not tend to act as a medium for the purchase, but as a means of providing information, which might not be the case for lower cost items. The likelihood of a negative outcome might be the same in both high cost and the low cost transactions, but the magnitude of loss will be greater in a higher cost transaction.

Another example to illustrate the importance of these aspects occurs when engaging in a business interaction. For example, where there is a 20% chance of failure that would result in a loss of $10, one might quite possibly engage in such an interaction. If, on the other hand, the loss from such an interaction were $200,000, one would probably not engage at a 20% failure rate but would require a much lower failure rate of say 0.01%, before being prepared to engage in such an interaction. The importance of each such representation is shown in Figure 2.1. Hence, it is not just the probability of failure alone, but the probability of failure taken together with the impact or consequences that guides the decision to participate in a business association. Such aspects are characterized by the notion of transactional risk. So, when transactional risk is assessed and analyzed in a business activity, it is imperative that it is done according to these aspects and the object of analysis.

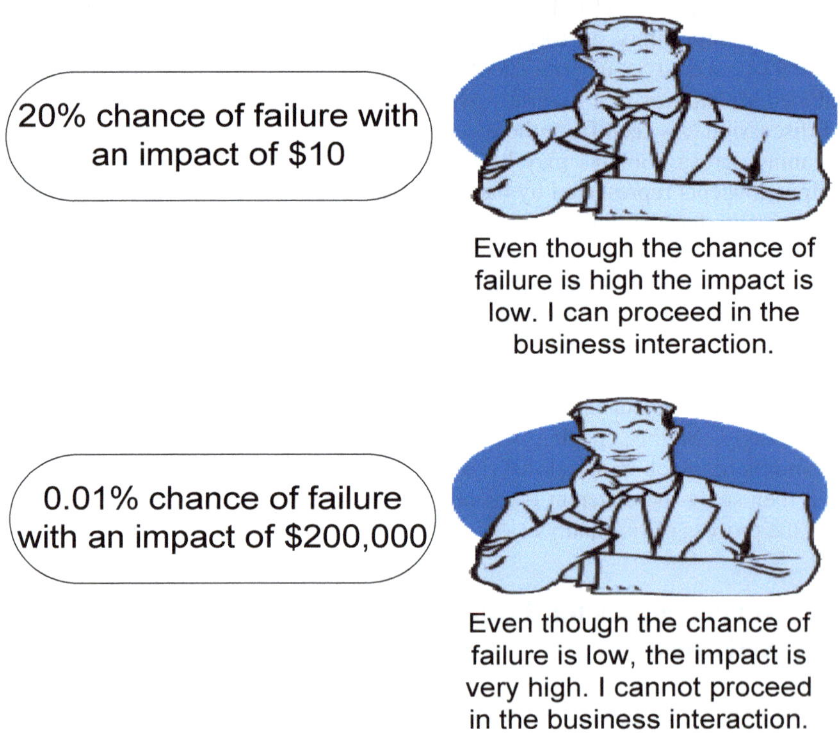

Fig. 2.1 Importance of probability of failure and impact in transactional risk analysis

Most approaches in the literature analyze transactional risk in business interactions as a sole measure of a single aspect of the outcome. In other words, they consider transactional risk in the business activity as either the probability of not achieving the desired outcomes of the interaction due to the associated threats [2-7], or the impact as a result of that [8-11]; but fewer approaches do it as a combination of both. However the quantification of only the threats associated in the business activity represents only one aspect of transactional risk, which is the probability that the interaction might fail and ignores the consideration of the other aspect, which is the impact as a result of that. Similarly, considering just the 'outcomes' or 'consequences' in the interaction represents only the aspect of financial risk but ignores the consideration of the other aspect, which is the probability of failure of the interaction. Such single aspect representations of transactional risk do not capture all the variables or factors required for its analysis, and hence, do not give the complete conclusion that is required for making an informed interaction-based decision. Therefore, to have such an informed representation of transactional risk, we propose that the specific aspects that should be analyzed in the context of business interactions for decision making are: (a) performance risk; and (b) financial risk as shown in Figure 2.2. These are discussed below:

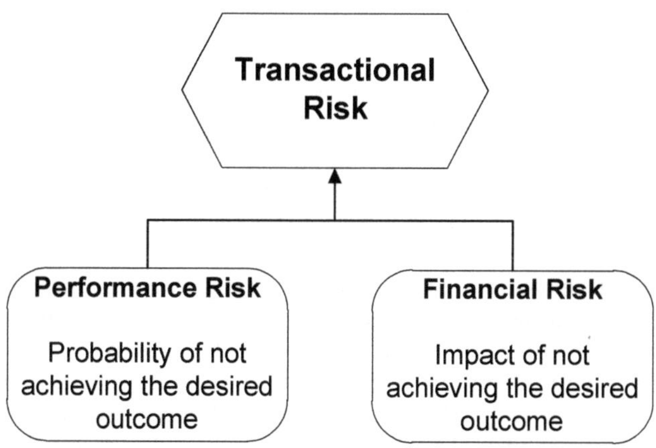

Fig. 2.2 Transactional risk and its sub-categories

2.2.1 *Performance Risk*

Performance risk represents the likelihood of the user not achieving the desired outcomes in the business interaction which he initiated. This may result in a negative outcome in the interaction. Such a likelihood may be related to various factors such as the inability or the willingness of the other user in not giving the interaction-initiating user the desired outcomes or other uncertainties which might lead to the non-achievement of desired outcomes.

2.2.2 Financial Risk

Financial risk represents the impact or the magnitude of loss which the interaction-initiating user will experience as the result of not achieving his desired outcomes in the business activity. The loss can be quantified in many different ways, but as a business activity involves financial resources, we consider that the impact will be in the loss (either total or partial) of these financial resources as a result of the non-achievement of desired outcomes. As is the case with performance risk, financial risk can be experienced due to various factors such as the inability or lack of willingness of the other user to give the interaction-initiating user desired outcomes or the occurrence of an event which will lead to the interaction-initiating user not achieving his outcomes desired from the interaction.

We will discuss these two sub-categories of transactional risk further and the process to analyze them in the next chapters. In the next section, we discuss about the two types of transactional risk.

2.3 Types of Transactional Risk

Transactional risk will have different meanings according to the timeframe in which it is being discussed. To illustrate with an example, let us consider the following two statements:

Statement 1: The management team wants to ascertain the risks which were ignored during the development of the project that led to its failure; and

Statement 2: Before starting the next project, the management team wants to ascertain and analyze the risks in its successful completion.

Both statements mention identifying or becoming aware of the possible threats, hazards and loss in the activity. But the first sentence deals with identifying the risks in the interaction '*after*' the occurrence of the activity, while the second sentence deals with identifying the possible risks in the interaction '*before*' the occurrence of the activity. The outcomes to which these risks will relate are very different from each other. In the first sentence, by identifying the transactional risk, steps can be taken to provide risk education or risk avoidance for future activities, as the activity in which the transactional risk is being assessed has already occurred; whereas, by identifying the transactional risk in the second sentence, the project management team can execute or take steps by which they can avoid or mitigate any possible loss due to the failure of the project, as the project is yet to start. To make an informed interaction-based decision in a business activity, the analysis of both types of transactional risk is important at different stages, as shown in Figure 2.3. Before an interaction is formed, information on both the past interaction and the current interaction are important for decision making, whereas after the interaction starts, the past interaction history of the user will have less bearing compared to the current analysis or strategies formed.

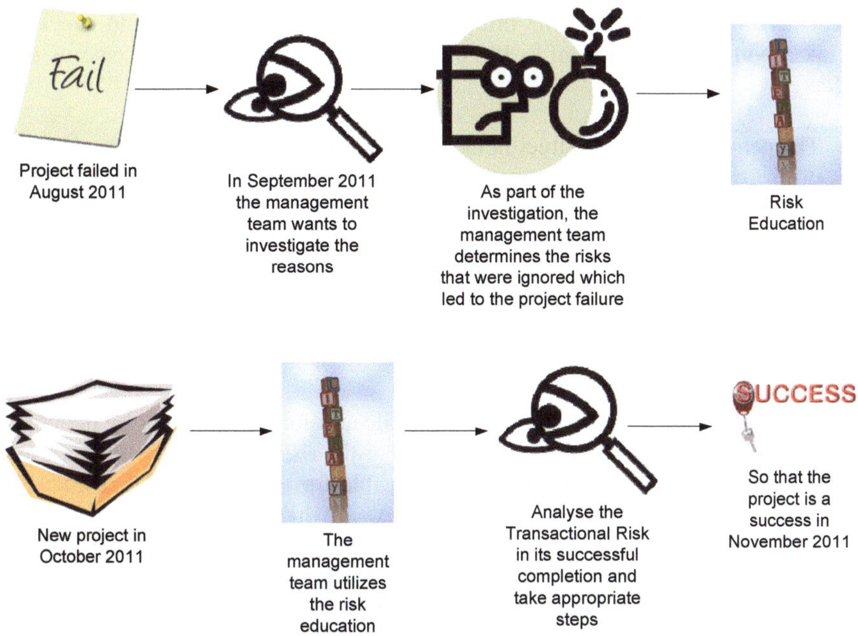

Fig. 2.3 Different types of transactional risk in the activity

In this book, we propose an approach by which an interacting user can determine beforehand the level of transactional risk in his business endeavor with another user/s. We interchangeably use the term *perceived transactional risk* to *transactional risk* and vice versa represent this type of transactional risk in the business activity. Further, in this book we consider that whenever perceived transactional risk is being analyzed for business decision-making, it is done according to:

(a) the subcategories of performance risk and financial risk, and
(b) its characteristics, which are explained in the next section.

2.4 Characteristics of Transactional Risk

Characteristics are the different properties on which perceived transactional risk in an activity is dependent. These properties take into consideration the dynamic and context specific nature of transactional risk which varies from time to time and from activity to activity. In other words, the characteristics of transactional risk are those factors that take into consideration the specific goal, purpose and time of the business activity when it is being analyzed for perceived transactional risk. The importance of these factors is illustrated by the following example and shown in Figure 2.4.

A person drives a car while drunk in an area where police plan to check for drink driving from 10:15 pm to 10:30 pm. The following possible scenarios may occur:

1. The person drives on a street in the same area in which police are checking for drink driving from 10:15 pm to 10:30 pm.
2. The person drives on a street in the same area in which police are checking for drink driving but after 10:30 pm.
3. The person does not drive at all after he is drunk.
4. The person is drunk but is a passenger in the car that goes through the same area in which police are checking for drink driving from 10:15 pm to 10:30 pm.

In discussion of the event 'person losing his license for being caught by the police while drunk', a careful analysis illustrates that the risk associated with this event varies for each scenario. The factors responsible for the variation of risk are the important details like the time and the context in which the person in question is being discussed in each event. In other words, the perceived risk in each scenario is different because the characteristics from which each event is being looked at varies in each of them. This leads to a change in their inter-dependencies and the impact that they might have on the object of analysis. So, it is important to consider such characteristics when perceived risk is being analyzed in a business activity. We discuss this further in the next sub-sections.

2.4.1 Context Specific Nature

Context defines the purpose for which the interaction is being carried out. During risk analysis, context specific nature takes into consideration the high level nature of the description of the business activity. For example in the discussion above, the context of the discussion can be *person losing his license after being caught by the police*. Similarly, as seen in the discussion of Section 1.2.2, the context of the discussion can be *manufacturer not meeting the customer's requirements.*

2.4.2 Criteria Specific Nature

Criteria are defined as the related functionalities or activities that come under a particular context. In the context of the business activity, there are different possible criteria. When a user carries out a business activity in a context, then it is possible that he may want to achieve only certain criteria in that context and not all the criteria that come under it. Similarly, when the perceived transactional risk in the business activity is being analyzed, it should be done according to the specific criteria and not the context in general. For example, in the discussion above, there may be many reasons why the person may be caught by the police. Some of these are: his car is below the road worthy level; he is driving an unregistered car; he is driving with 4 people seated in the back of the car when the maximum capacity is 3, etc. But, in the above scenario, the risk is being assessed in relation to the person being caught by the police as a result of drink driving. So, it should be analyzed only in accordance with this. Similarly, in Section 1.2.2, the context is the manufacturer not meeting the customer's requirements due to factors such as failure of production units, non-availability of raw materials etc. So, when risk is being determined, it should take into consideration only those factors such as failure of production units, non-availability of raw materials etc.

2.4.3 Dynamic Nature

Dynamic nature represents the variability of transactional risk according to the specific time period. For example, in the discussion of Section 2.4, the police plan to check for drink driving between 10:15 pm to 10:30 pm. A person may be caught if he is drunk and drives along that route during that time period. If the person drives at a different time period, then the risk of him being caught may be less as the police may not be preset during that time. So, risk is dynamic and time plays an important part in this analysis. The transactional risk which a user has to analyze could be in a set of business activities with the other user/s over a given finite period of time, or might be in a single activity which extends over that finite period of time. Furthermore, the time period of the initiating-user's business activity with the other users might extend to a point in time in the future, subsequently influencing his behavior in forming an interaction by the outcome of the risk analysis over that finite period of time. So, when the perceived transactional risk in the business activity is being analyzed, it should be modeled accurately according to the period of time in which it is being determined in the interaction. Further explanation about the dynamic nature of risk is discussed in the next chapter.

In the next section, we will discuss the different types of uncertain events that need to be identified and then analyzed during perceived transactional risk analysis.

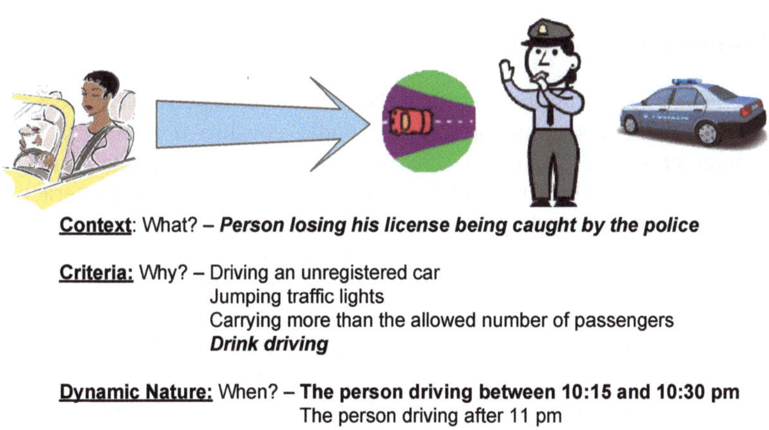

Fig. 2.4 Different characteristics of transactional risk in an activity

2.5 Types of Business Activity

As discussed earlier, perceived transactional risk in business activities represents the occurrence of undesired events and their associated consequences. These undesired events are from the perspective of the interaction-initiating user. But from whom does the interaction-initiating user want to achieve these outcomes? The answer to this question varies according to the purpose or scenario for which the activity is being carried out and can be classified into two sub-types as discussed next.

2.5.1 Single User Activity

Some activities are such that they are dependent only on the interacting user to achieve a successful outcome. An example of single user activity is 'a person losing his license if he is caught driving while drunk'. If the person in question does not drive when he is drunk, then he will not get caught and will not lose his license. In other words, as shown in Figure 2.5, the interacting user alone is responsible for having a successful outcome or for avoiding a negative outcome in such an activity. This might not be the case in all activities which will lead to the second type of activity as explained in the next sub-section.

Fig. 2.5 Single User Activity

2.5.2 Collaboration-Based User Activity

Other activities are such that there needs to be collaboration between at least two users to achieve a successful outcome. For example, consider a business activity where an interacting user wants to achieve a desired outcome, like buying a Playstation 3 from the seller for \$500. In such a business activity, the interacting user 'A' is not solely responsible for achieving a successful outcome, rather there needs to be cooperation from the seller as well. A possible scenario in such an interaction is that the buyer pays the seller \$500 but the seller may not send the Playstation 3 to the buyer. This may result in the occurrence of a negative outcome and subsequently loss to the interacting user. To have successful completion in the business activity, the interacting user wants the seller to cooperate and send the Playstation 3 to him upon the receipt of payment, as shown in Figure 2.6.

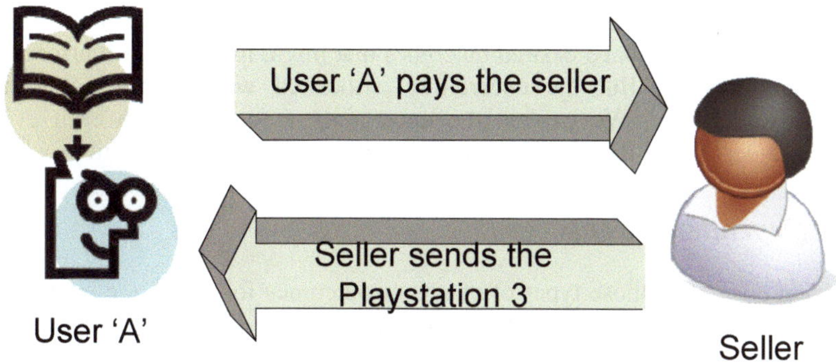

Fig. 2.6 Collaboration-based User Activity

So, the source from which the interacting user wants to achieve his desired outcomes for the successful completion of the business activity varies according to the interaction scenario. In this book, we consider that the interacting user carries out a business activity, which in turn, needs cooperation and collaboration between at least two users for the successful completion of the activity. In other words, the events that the interaction-initiating user wants to achieve for the successful completion of the business activity are dependent on other user/s. Subsequently, the uncertainty associated when analyzing perceived transactional risk represents the fuzziness of the interacting user in having a definite conclusion of the other user/s actually committing to the criteria or events as desired in the business activity. But not all events of the business collaboration depend on the other user. There might be some events that are not dependent on the other user but still have uncertainty associated with them. Identifying and analyzing these is important during transactional risk analysis. We discuss this further in the next section.

2.6 Types of Events Related to Transactional Risk in Business Activities

To discuss the different types of events related to transactional risk in business activities, let us consider an example of an interacting user 'A' wanting to interact with a logistics company in order to achieve the following desired outcomes:

- The logistics company should pack the goods properly at the pick up address **(Desired Outcome 1).**
- The goods should reach the destination in 5 days **(Desired Outcome 2).**
- The goods should reach the destination undamaged **(Desired Outcome 3).**
- The logistics company should provide a track and trace facility for the user **(Desired Outcome 4).**
- The total cost for transferring the goods is 3000 euros. User 'A' should pay half of the amount when the goods are picked up **(Desired Outcome 5).**
- Once the goods are delivered, user 'A' should make the remaining payment of 1500 euros in Australian dollars to the logistic company's branch office in Perth **(Desired Outcome 6).**

For transactional risk analysis, user 'A' has to identify the uncertainty in achieving each of the abovementioned desired outcomes that might lead to the occurrence of a negative outcome. But the uncertainty associated with each event varies according to its type. We explain the uncertainty associated with each event further after describing the two different types of events in the next section.

2.6.1 Dependent Events

Dependent events are those types of events or outcomes for which the interacting user depends on the cooperation of the other user/s in the interaction, for example *Desired Outcomes 1-4* in above section. Such dependent events are desirable events which the interacting user wants to achieve to have at least a neutral outcome in the activity, as shown in Figure 2.7.

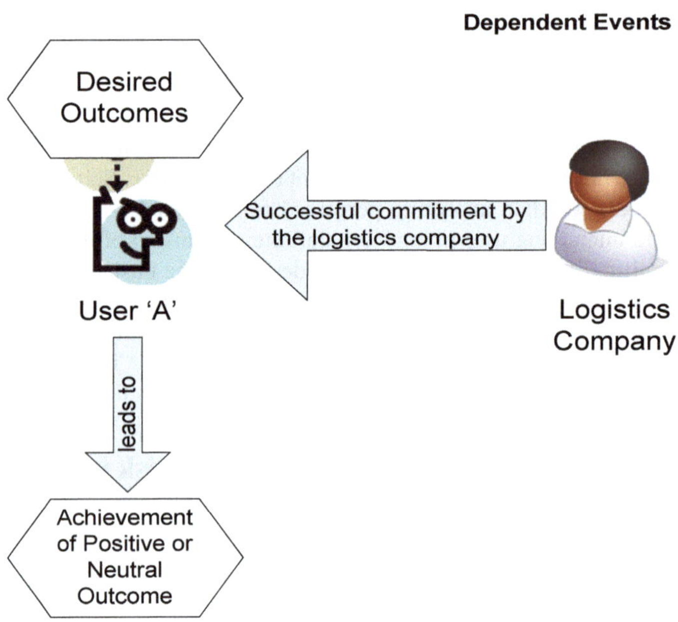

Fig. 2.7 Dependent Events on the Logistics Company

2.6.2 Non-dependent Events

Non-dependent events are those events that are outside the scope of dependence on the other user, for example *Desired Outcomes 5 and 6*. Such non-dependent events need to occur in the desired way to achieve at least a neutral outcome in the activity, as shown in Figure 2.8.

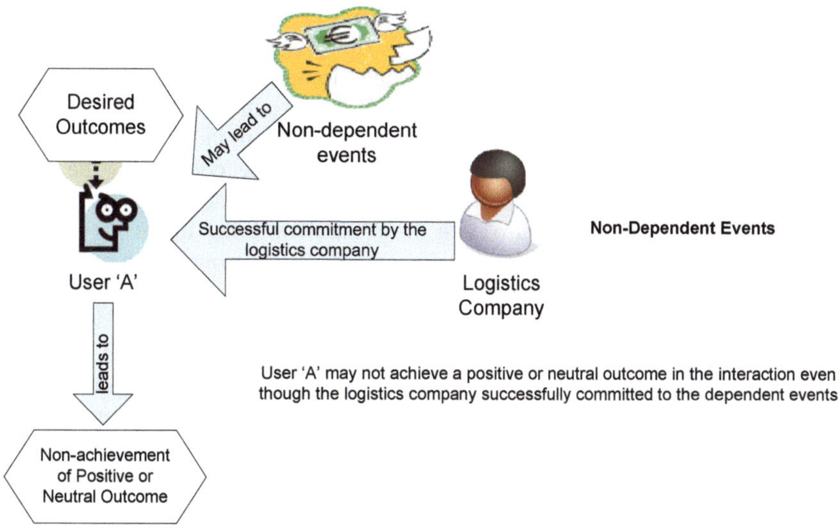

Fig. 2.8 Non-Dependent Events on the Logistics Company

2.7 Types of Uncertainty Related to Transactional Risk in Business Activities

According to the type of event, during risk analysis the uncertainty associated with each of these types of events is different. Let us consider that user 'A' is ready to commit to *Desired Outcomes 5* and *6* that require him to pay the required financial resources at different time periods. Then, to analyze the transactional risk and the likelihood of him not achieving his desired outcomes in the business activity, he will only have to analyze the uncertainty of the logistics company in committing to his part of the collaboration. But apart from analyzing this type of uncertainty, there is another type of uncertainty related to the collaboration that is very important for interacting user 'A' to consider while carrying out risk analysis. This type of uncertainty, too, will have a significant part to play to achieve the successful completion of a business activity. We discuss the uncertainties associated with them in the following section.

2.7.1 Uncertainty Associated with Dependent Events

Failure to achieve the dependent criteria or events in accordance with user 'A's desired outcomes might result in the user experiencing failure in his business activity along with experiencing a negative impact. The uncertainly associated with these types of events is that user 'A' wants the logistics company to commit to the collaboration but he is uncertain if this will be the case. Such uncertainty can be alleviated by determining the level of capability of the logistics company in committing to the desired events.

2.7.2 Uncertainty Associated with Non-dependent Events

Failure to achieve the non-dependent events in accordance with the user's expectations, too, will result in him experiencing failure in their business activity. In other words, these events, too, will have either a direct or indirect impact on the successful completion of the business activity. For example, considering the above discussion of user 'A's business activity with the logistics company, it is possible that the exchange rate of the euro with the Australian dollar might vary, which will result in user 'A' paying more than the amount that was agreed in outcome 6. As shown in Figure 2.8, from agent 'A's point of view, the occurrence of such a scenario in this event can be termed as undesirable, and has a chance of taking the business interaction in a negative direction. The uncertainty associated with this type of event is that user 'A' is unsure as to whether it will occur in the interaction, according to his expectations, so that at least a neutral outcome is achieved. Variation in the occurrence of such non-dependent undesired outcomes may be volatile and difficult to analyze beforehand, according to a certain pattern or history, as can be done with dependent events. But the financial loss that can be experienced from such non-dependent events is different from what could be experienced from dependent events, and subsequently such events, too, should be identified and their uncertainty assessed during transactional risk analysis in a business activity. This is discussed further in Chapter 6.

So, while carrying out transactional risk analysis, the interacting user should consider all the events that are related to his business activity and then determine the type of uncertainty or uncertainties that are associated with each of them, according to the type of event. Once these have been identified, then each subcategory of perceived transactional risk should be determined according to its characteristics to have a complete analysis in business interactions. In the next section, we will summarize the issues in the literature to determine the level of perceived transactional risk in the business activity, according to its subcategories and characteristics.

2.8 Issues in Ascertaining Perceived Transactional Risk in the Literature

In this section, we discuss briefly the drawbacks of the existing approaches in the literature or the issues involved while analyzing perceived transactional risk.

2.8.1 Summary of Approaches to Determine Performance Risk

In business interactions, the performance risk of a user is determined either by his past interaction history or/and by his reputation. Different mathematical frameworks to model the reputation of a user have been proposed in the literature. Some of these approaches utilize the global reputation of the user [12-16], but utilizing the global reputation results in losing or omitting the context-specific nature and the criteria-specific nature of the current interaction against which the performance risk has to be determined. To avoid this, approaches have been proposed which consider the context-specific nature of the interaction by taking care to solicit recommendations only in the context in which the performance risk is to be determined. But they do not consider the criteria-specific and dynamic nature of risk while determining the reputation of an agent [2, 17-19]. An agent making use of such a context-specific nature to determine the performance risk is basing his decision on values which 'may be' related to this interaction, without the dynamic behavior of the agent. We use the term 'may be' because, as discussed earlier, in a given context, there might be a number of criteria, and if an interaction-initiating user is interacting with another user in a particular context, this does not mean that he wants to achieve all of the criteria in that context, as shown in Figure 2.4. Rather, he might want to achieve only certain criteria in that context which forms the basis for the desired outcomes of his interaction. Hence, even the approaches proposed in the literature which consider the context specific nature of the interaction while determining the reputation of an agent, might not ascertain the accurate performance risk in the interaction as it may have different criteria in that context. Some approaches in the literature consider the context and temporal aspect or time while assessing the perceived risk in the transaction, but they omit to consider this along with the criteria-specific nature of the interaction [20]. The interaction-initiating user, making use of such a finite aspect of time, may consider the dynamic property of risk, but the performance risk that he determines may not be in the specific context and criteria of his future interaction, and the omission of these aspects will result in determining a level of performance risk which might be unrelated to the desired outcomes of his future interaction. Similar arguments hold for the approaches proposed in the literature which model the performance risk by using the past interaction history of an agent.

Further, reputation or trustworthiness determines and represents the capability of an agent to commit to the desired outcomes. To determine the performance risk, which shows the inability of the user to commit to the desired outcomes of the interaction-initiating user by highlighting his vulnerabilities, researchers determine this as 1-reputation or 1-trust and represent it as the possible 'risk' in the interaction [21]. This gives rise to a major misconception in the literature, which terms the perceived risk in the interaction as the complement of trust, and subsequently a measure of just one of its subcategories. Furthermore, none of the researchers consider the context specific, criteria specific and dynamic nature of the interaction when determining the trustworthiness or/and reputation of an agent, and subsequently the 'risk' which they determine from it is not according to those characteristics of its interaction. Moreover, even if the characteristics of the interaction were being considered when determining the possible 'risk', then it at most represents one subcategory of perceived risk, namely performance risk, ignoring its other subcategory, financial risk, which is a crucial consideration when determining the perceived risk in the context of business interactions.

2.8.2 Summary of Approaches to Determine Financial Risk

Financial risk in business interactions represents the magnitude of loss that could be experienced as a result of risk. In determining financial risk, no approach has been proposed in the literature which determines the degree and level of financial loss in the interaction, according to the level of performance risk of the user. Some approaches place the transaction amounts into different categories, and then determine whether or not to risk that particular amount based on the level of trust in the agent [21]. These types of approaches in no way determine and represent explicitly the financial risk of the financial loss that can be experienced in the interaction; rather, they merely utilize the financial aspect of the interaction and determine the impact of the level of trust in the interaction on it, and term it as risk to utilize it in decision making, a misconception which is also evident while determining performance risk. Other approaches consider the utility functions to identify the risk [14], the level of gain in the interaction [22], the cost of all outcomes [18] or cost of benefit of the outcome [23]. But, financial risk means identifying the possible financial loss in the interaction due to the threats posed in interacting with the other agent, and the current approaches do not represent the possible degree and level of financial loss that can be experienced in the interaction. Moreover, none of the approaches in the literature which consider the financial aspect of the interaction determine the financial risk according to the dynamic nature of investment in the different time scenarios of the interaction, even when it extends to a future point in time.

So, an approach is needed by which the level of perceived transactional risk is considered according to its subcategories and characteristics in a business activity. This is important for making an informed interaction-based decision in a business activity. In the next sections, we will discuss what needs to be done when transactional risk is determined in an interaction.

2.9 Risk Consideration in Business Activities

Once the level of perceived transactional risk has been determined, then the interaction-initiating user will have the following options to choose from:

1. Risk does not have any effect on me. Let's get started with the business activity.
2. The risks associated with the activity are within my tolerable limits.
3. The risks associated with the activity are too high. I need to see if I can possibly avoid, alleviate or mitigate them.
4. The risks associated with the activity are too high, even after taking steps to address them. There is no value in proceeding.

There are different types of users in the community and it is not surprising if their views fall into one of the above categories. But not all of them are so irrational as to reach the first conclusion in the face of significant risk. Most of us consider perceived transactional risk as a significant threat to the successful completion of a business interaction and conclude on an option from 2 to 4. But to reach any of these conclusions, the interacting user has to carry out a series of steps. These steps come under the broad category of risk consideration.

Risk consideration entails the broad category of steps that identifies the events which may lead to adverse effects, analyzes the probability of occurrence of those events, ascertain their impact, determines the level to which they can be tolerated and then treats and monitors them. As shown in Figure 2.9, this process of dealing with risk is a combination of two broad steps which we categorize into 'risk analysis' and 'risk management'.

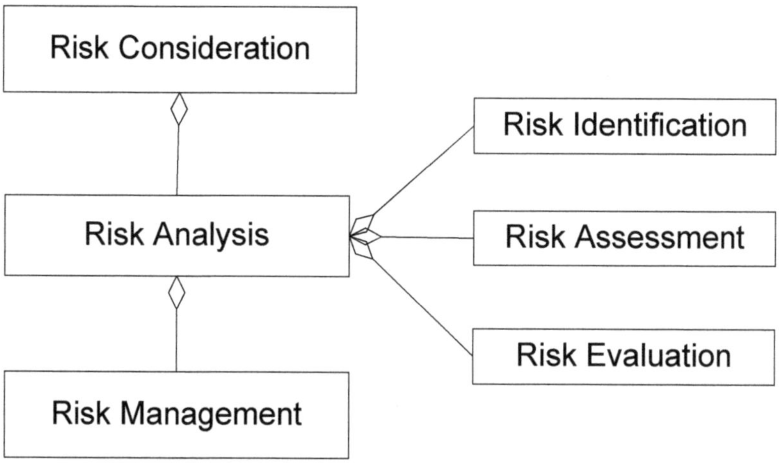

Fig. 2.9 Process of risk consideration in business activities

Each step has its own role to perform in making an informed interaction-based decision. Further, as shown in Figure 2.10 in each step, there are various sub-steps that need to be carried out to complete the process of risk consideration in business activities. We will give a brief explanation of each of these components and the sub-steps involved in them.

2.9.1 Risk Identification

The first step in risk consideration is risk identification. The primary aims of this step are to establish the context of the activity, identify its scope and then ascertain the different types of events that may have uncertainty associated with them. This will give the possible threats to the successful completion of its activity. It is important to note that the details of how these steps need to be carried out in an activity and the output to which they lead can vary with the field of discussion and the nature of the associated risks. The output of this step will lead to the next series of steps which will be discussed under risk assessment.

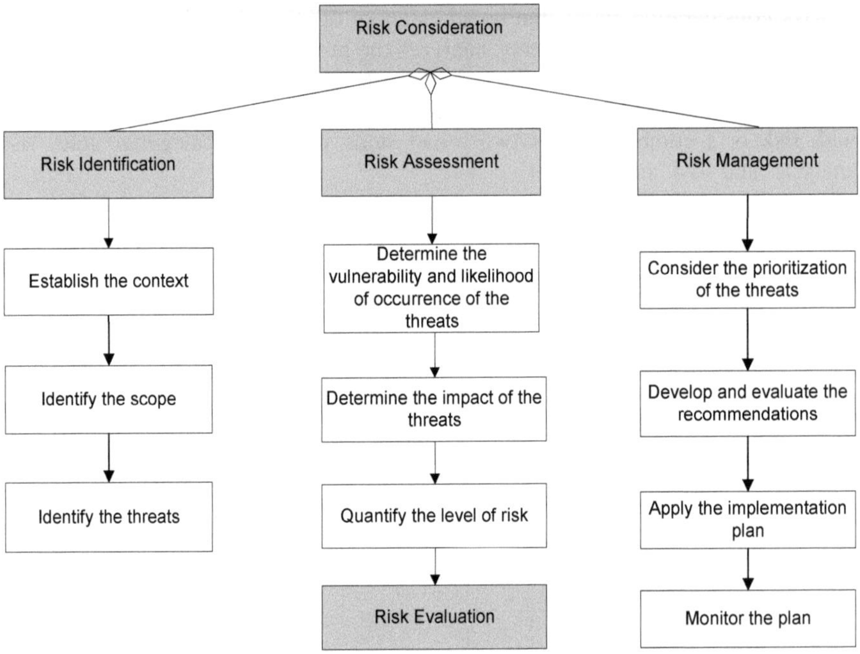

Fig. 2.10 Components and sub-steps for risk consideration

2.9.2 Risk Assessment

In any activity, risk assessment is an important step or a major hurdle to clear before any decision can be taken on the future course of action. This step gives an insight into the major potential threats and the severity of their impact on the successful completion of the business activity. This output of this step will form inputs to the step of risk evaluation.

2.9.3 Risk Evaluation

Based on the assessed risk in this previous step, the aims of this step are to (a) weigh these risks according to the risk attitude of the interacting user; and (b) if there are many risks, serve as the basis for prioritizing and resolving the identified risk events. The output of this step will lead to the next series of steps that provides the choice of exploring the alternative options, for managing, minimizing or avoiding risk. These series of steps come under risk management, and its aims are to deal with the identified risk events and the potential options to treat and address them.

2.9.4 Risk Management

The purpose of risk management is not to immediately avoid an activity that has risk associated with it by considering it as a threat to the successful completion, but rather, to carry out a series of steps by which the identified risk events in that activity may be alleviated or minimized to an acceptable level. Of course, risks are meant to be contained in a business activity, but they should be done so that if an event still remains and is classified as a risk, after all the steps have been taken to alleviate or minimize this risk. Risk identification, assessment, evaluation and management is a proactive process in which decisions have to be taken according to the severity of the identified risk events so that they are minimized or mitigated to a certain level. An important point to mention here is that the risk management process may not guarantee the complete elimination of the uncertainties of the risk events, but at most *reduces* or *minimizes* them in an interaction to a certain level that is termed as *acceptable* to the interacting user. This will help to achieve a higher probability of achieving either a neutral or positive outcome in the activity and hence is of considerable importance.

The process of risk identification, assessment, evaluation and management has to be carried out according to a series of standardized steps. The ISO standard on risk management details the different steps that need to be carried out in this process. In the next section, we briefly discuss this standard for risk management and the different steps by which risk should be identified and managed.

2.10 ISO Standard on Risk Management

Different standards on risk management have been proposed in the literature. These standards define the foundations of terms, definitions and framework for risk management that can, in turn, be used as a basis for decision-making and accountability. In the past, there were different standards proposed in the literature, but efforts are being made to have a single and unified standard for risk management. In this section, we will discuss the AS/NZS ISO 31000:2009 [24]. This standard is identical to the ISO 31000:2009 standard and follows its terminology. We will give an overview of the process as defined in that standard which is important for risk management.

2.10.1 Framework for Risk Management

The framework proposed by the ISO standard for risk management consists of five main components. These components follow an iterative cycle and are inter-related with each other. A brief explanation of the framework is given next. For a detailed explanation, readers should refer to [24].

➢ **Mandate and Commitment**

This component stresses the importance of having an ongoing and sustained commitment from the organizations' management as well as having strategic and rigorous planning for risk management. Some of the important factors in this component are: defining the risk management policy, developing the performance indicators, aligning the risk management objectives with that of the organization, having clear communication with stakeholders etc [24].

➢ **Framework for managing risk**

This is one of the key components of the framework and it defines various important factors such as identifying and establishing the current situation of the organization from an external and internal perspective, establishing the risk management policy, developing accountability measures, incorporating the risk management policy in the organization's day-to-day working and policies, and establishing communication policies etc.

➢ **Implementing risk management framework**

This component defines strategies which organizations should follow while implementing the framework and the process of risk management [24]. Some of the key factors while implementing the framework for risk management are: defining the appropriate timing and policies for implementing the framework; complying with the legal and regulatory framework; and educating the public and stakeholders on these policies etc.

➢ **Monitoring and reviewing the framework**

This component defines the steps to ensure that the process of risk management is effective, and monitors and reviews the framework to ensure that the framework stays current and performs its tasks. Some of the key tasks in this component are: measuring the risk management performance against indicators, commitment to risk management plan, and the timely update of the risk management framework etc [24].

➢ **Improvement of the framework**

Based on the analysis of the previous step, this component looks at the continual improvement of the risk management framework, policy and plan [24]. This will help an organization to improve its risk management culture.

The ISO standard also defines the various steps in the process of risk management. We explain this briefly in the next section. For a detailed explanation, readers should refer to [24].

2.10.2 Steps in the Risk Management Process

The process of risk management, as defined in the ISO standard, is a combination of six components. These are as follows.

➢ **Communication and Consultation**

This component stresses the importance of having a communication and consultation process with different stakeholders to understand their perception of risk and to have a common understanding. Other important steps in this component are identifying the consequences and the measures taken to treat risk etc. This is a key step in developing plans for identifying and mitigating risk at an early stage.

➢ **Establishing the context**

This component establishes the context from both an external and internal perspective of the organization. This is an extension of the framework for managing the risk component defined earlier, and it identifies in more detail the various factors related to the risk management process from contexts both external and internal to the organization. The external context will consider factors from aspects such as social and cultural, political, legal etc whereas the internal context considers the objectives and criteria of the particular project, the organizational culture and the values of the stakeholders. This will lead to the next component of risk assessment which includes the steps of risk identification, risk analysis and risk evaluation.

➢ **Risk Assessment**

This component details the overall process of risk identification, analysis and evaluation. Risk identification deals with identifying the different sources of risk, areas of impact, events and their possible impacts. Risk analysis deals with the analysis of the different risk events, determination of likelihood of occurrence and the consequences to which they lead. This will lead to the next step of risk evaluation. Based on the assessed level of risk, this step evaluates them and assists while decision making. Further, this step will also help to determine which risks need treatment and at what level of priority.

➢ **Risk treatment**

Risk treatment deals with selecting, preparing and choosing appropriate strategies for treating the identified risk events.

➢ **Monitoring and review**

This step deals with developing a plan that continuously monitors emerging risks, detects any changes to the external and internal factors and utilizes this information to improve the process of risk assessment.

➢ **Recording the risk management process**

In this step, the processes taken for risk management are recorded so that they can be utilized for future education and used for the improvement of methods and tools.

These series of steps should be followed while carrying out the process of risk consideration. Risk consideration has become an integral part of various disciplines including finance, health, business transactions, safety regulations, workplace safety environments etc. as it helps to avoid undesirable outcomes by assessing and managing the perceived risk in the transaction. But, as discussed earlier, events that are possibly identified as risk in an activity may not be considered as risk in any other activity, or will not have the same level of negative effect on the other activity. Subsequently, the way by which the interacting user sees such events in these two types of interactions varies, consequently changing the process of risk consideration in each of them. So, as with risk assessment, the process of risk management, too, varies across each interaction discipline and the user for whom it is carried out. Hence, before carrying out the process of risk consideration, it is important to understand the interaction according to the activity's perspective. In the next section, we will discuss the process of carrying out risk management in business activities.

2.11 Ways of Managing Risk in Business Activities

The development of the Internet has given its users more flexibility and various functionalities which facilitate or influence the way business interactions are carried out. But, with the provision of increased functionality also comes the fear of loss or the fear of not achieving what is desired in an interaction. To achieve a successful outcome in the business activity, such fear of loss has to be determined and in most cases, reduced to a certain acceptable level which will, in turn encourage the user to form a business interaction. This is achieved by risk assessment and management. However, as discussed earlier, the process and the sequence of steps for risk assessment and risk management are specific to the activity in which they are being carried out. So, it is important for the interacting user to first understand how risk is categorized in the activity before carrying out such a process. As mentioned in the next sub-section, there are two broad ways by which the process of risk management can be carried out, according to the domain of the activity. These are qualitative and quantitative.

2.11.1 Qualitative Process

Qualitative approaches mainly stress the importance of establishing a dialogue among the different stakeholders or participants to identify and categorize the different risks according to the risk matrix or risk calculator. The effect of those risks is then determined on the consequences scale that is classified into different categories. Based on the determined consequences, strategies are developed by which risks are managed or treated and monitored continuously.

2.11.2 Quantitative Process

Quantitative approaches mainly deal with numerical techniques that ascertain the severity of the risk events, the probability of their occurrence according to a certain past pattern of occurrence and then determine their impact on the successful completion of the activity. This will then lead to the process of classifying the different events, according to their severity and then carrying out steps to mitigate them.

The qualitative process of risk management is best suited and utilized in domains such as the construction industry or managing operational risks in the running of a chemical plant. In such scenarios, identifying the different risks, determining their consequences on a linguistic scale, and then developing strategies according to their severity by which they can be monitored and treated continuously is the best process to manage them. Such qualitative risk assessment and management processes will not be best suited to the domain of business activities. Moreover, it is highly possible that stakeholders in business interactions are usually the people initiating the business activity and not an extended large group, as with other domains. So, in such domains, a systematic and quantitative approach is needed by which the different identified risks are assessed for their magnitude and their consequences or impact in order to manage them. But, no matter what approach for risk management is chosen, the steps to be followed should conform to the basic steps of the ISO international standard on risk management.

As our focus in this book is on transactional risk assessment in business interactions, in the next chapter, we will propose an approach by which it can be assessed and analyzed in that domain. We begin by first defining perceived transactional risk, according to its object of analysis in business activities and then propose the framework by which it can assessed.

2.12 Recapitulation

In this chapter, we discussed the subcategories and characteristics according to which perceived transactional risk should be analyzed for making informed interaction-based decisions in business activities. We then discussed the different types of perceived transactional risk and uncertain events that need to be considered during its analysis. The issues in the literature relating to the analysis of perceived transactional risk, according to its subcategories and characteristics were summarized. We then discussed the process of risk consideration in business interactions and briefly looked at the ISO standard on risk management. In the next chapter, we will introduce our framework for transactional risk analysis in business activities.

References

1. Sander Greenland: Bounding Analysis as an Inadequately Specified Methodology. Risk Analysis 24, 1071–1083 (2004)
2. Gran, B.A., Fredriksen, R., Thunem, A.P.-J.: An Approach for Model-Based Risk Assessment. In: Heisel, M., Liggesmeyer, P., Wittmann, S. (eds.) SAFECOMP 2004. LNCS, vol. 3219, pp. 311–324. Springer, Heidelberg (2004)
3. Griffiths, N.: Task Delegation using Experience Based MultiDimensional Trust. In: Proceedings of the 4th International Conference on Autonomous Agents and Multiagent Systems (AAMAS 2005), pp. 489–496. ACM Press, Utrecht (2005)
4. Corbitt, B.J., Thanasankit, T., Yi, H.: Trust and e-commerce: a study of consumer perceptions. Electronic Commerce Research and Applications 2, 203–215 (2003)
5. Wojcik, M., Eloff, J.H.P., Venter, H. S.: Trust Model Architecture: Defining Prejudice by Learning. In: Fischer-Hübner, S., Furnell, S., Lambrinoudakis, C. (eds.) TrustBus 2006. LNCS, vol. 4083, pp. 182–191. Springer, Heidelberg (2006)
6. Xu, J., Liu, Z., Li, Y.: Integrating Processes of Logistics Outsourcing Risk Management in e-Business. In: IEEE/WIC/ACM International Joint Conference on Web Intelligence and Intelligent Agent Technology, pp. 544–547. IEEE, Hong Kong (2006)
7. Castelfranchi, C., Falcone, R.: Trust Is Much More than Subjective Probability: Mental Components and Sources of Trust. In: Proceedings of the 33rd Hawaii International Conference on System Sciences, pp. 1–10. IEEE Computer Society, Hawaii (2000)
8. Cvrček, D., Moody, K.: Combining Trust and Risk to Reduce the Cost of Attacks. In: Herrmann, P., Issarny, V., Shiu, S.C.K. (eds.) iTrust 2005. LNCS, vol. 3477, pp. 372–383. Springer, Heidelberg (2005)
9. Bohnet, I., Zeckhauser, R.: Trust, risk and betrayal. Journal of Economic Behavior & Organization 55, 467–484 (2004)
10. Xie, G., Zhang, J.-L., Lai, K.K.: Web-Based Risk Avoidance Group Decision Support System in Software Project Bidding. In: International Conference on Web Intelligence and Intelligent Agent Technology, pp. 180–183. IEEE, Hong Kong (2006)
11. English, C., Wagealla, W., Nixon, P., Terzis, S., Lowe, H., McGettrick, A.: Trusting Collaboration in Global Computing Systems. In: Nixon, P., Terzis, S. (eds.) iTrust 2003. LNCS, vol. 2692, pp. 136–149. Springer, Heidelberg (2003)
12. Su, C., Zhang, H., Bi, F.-M.: P2P-based Trust Model for E-Commerce. In: IEEE International Conference on e-Business Engineering (ICEBE 2006), pp. 118–122. IEEE Computer Society, Shanghai (2006)
13. Lin, C., Varadharajan, V.: Trust Based Risk Management for Distributed System Security - A New Approach. In: The First International Conference on Availability, Reliability and Security (ARES 2006), pp. 6–13. IEEE Computer Society, Vienna (2006)
14. Lam, Y.-H., Zhang, Z., Ong, K.-L.: Trading in Open Marketplace Using Trust and Risk. In: Proceedings of the IEEE/WIC/ACM International Conference on Intelligent Agent Technology (WI-IAT 2005), pp. 471–474. IEEE, Compiegne (2005)
15. Deriaz, M.: Trust without Truth. In: Etallc, S., Marsh, S. (eds.) Trust Management. IFIP AICT, vol. 238, pp. 31–45. Springer, Boston (2007)

16. Singh, A., Liu, L.: TrustMe: Anonymous Management of Trust Relationships in Decentralized P2P Systems. In: Third International Conference on Peer-to-Peer Computing, pp. 142–149. IEEE Computer Society, Linköping (2003)
17. Asnar, Y., Giorgini, P., Mulyanto, A.: Risk in Secure and Dependable System: a Survey. In: Proceeding of 1st International Conference and Workshop on Risk Technology and Management, Bandung, Indonesia, pp. 1–12 (2007)
18. Dimmock, N., Belokosztolszki, A., Eyers, D., Bacon, J., Moody, K.: Using trust and risk in role-based access control policies. In: Ninth ACM Symposium on Access control Models and Technologies, New York, pp. 156–162 (2004)
19. Wawrzyniak, D.: Information Security Risk Assessment Model for Risk Management. In: Fischer-Hübner, S., Furnell, S., Lambrinoudakis, C. (eds.) TrustBus 2006. LNCS, vol. 4083, pp. 21–30. Springer, Heidelberg (2006)
20. Xiong, L., Liu, L.: A Reputation-Based Trust Model for Peer-to-Peer eCommerce Communities. In: IEEE International Conference on E-Commerce, California, USA, pp. 275–284 (2003)
21. Wang, Y., Wong, D.S., Lin, K.-J., Varadharajan, V.: Evaluating transaction trust and risk levels in peer-to-peer e-commerce environments. Information Systems and E-Business Management 6, 25–48 (2008)
22. Jøsang, A., Presti, S.L.: Analysing the Relationship between Risk and Trust. In: Jensen, C., Poslad, S., Dimitrakos, T. (eds.) iTrust 2004. LNCS, vol. 2995, pp. 135–145. Springer, Heidelberg (2004)
23. English, C., Terzis, S., Wagealla, W.: Engineering Trust Based Collaborations in a Global Computing Environment. In: Jensen, C., Poslad, S., Dimitrakos, T. (eds.) iTrust 2004. LNCS, vol. 2995, pp. 120–134. Springer, Heidelberg (2004)
24. Standards Australia: AS/NZS ISO 31000:2009 Risk Management - Principles and guidelines. In: Standards Australia (ed.), pp. 1–29 (2009)

Chapter 3
Framework for Transactional Risk Analysis in Business Activities

3.1 Introduction

In the previous chapter, we discussed the various characteristics of transactional risk. We also noted that the way transactional risk is defined and interpreted across various domains varies. As such, the representation and assessment methods of transactional risk in any domain cannot be used to define and assess it in other domains, as the object of analysis across domains changes. So, to have an effective framework for transactional risk analysis in business activities, it is imperative to have a clear understanding of how transactional risk is categorized and understood in that domain. In this chapter, we will introduce our definition of perceived transactional risk in the domain of business activities and propose a framework by which it can be analyzed, according to the characteristics and object of analysis.

3.2 Definition of Key Concepts for Transactional Risk Analysis in Business Activities

Before we present our definition, in this section we provide an understanding of the concepts that we are going to use to define perceived transactional risk in business activities. Some of the concepts discussed below have been defined by other researchers in the literature and sometimes these tend to be different. To have a common and unified understanding of the key concepts, we define them according to how we use them from here on, in this book.

3.2.1 Agent

An 'agent' is defined as a specialized intelligent entity which acts autonomously according to the scenario and situation in which he currently is, in order to achieve his specified objectives. An agent can be either a software agent or a human agent.

O.K. Hussain et al.: Risk Assessment & Management in the Networked Economy, SCI 412, pp. 55–83.
springerlink.com

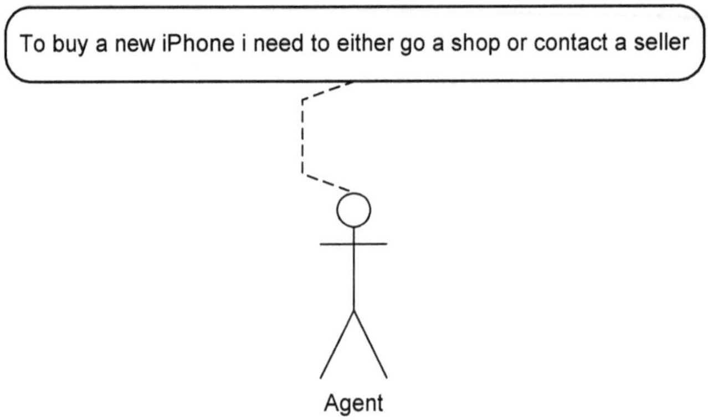

Fig. 3.1 Representation of an agent

3.2.2 Risk Assessing Agent

'Risk assessing agent' is defined as the interaction-initiating agent of the business activity. This agent wants to achieve certain outcomes that will lead him to initiate either a single user or a collaboration-based activity. The risk assessing agent is also the agent who analyses the possible perceived transactional risk present in interacting with another agent. So at a generic level, the risk assessing agent is defined as the decision-making agent who assesses the perceived transactional risk in initiating a business interaction.

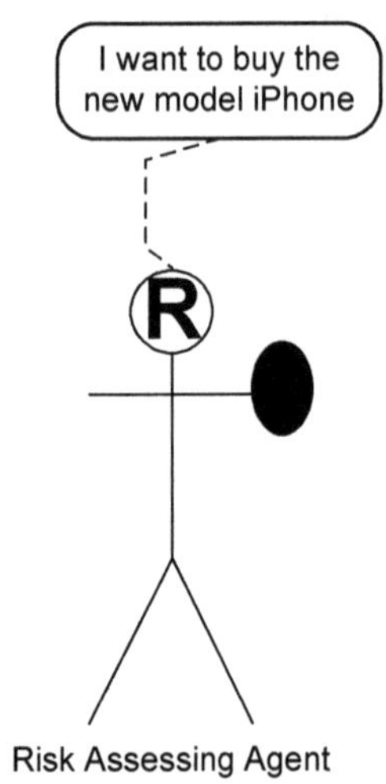

Fig. 3.2 Risk assessing agent

3.2.3 Risk Assessed Agent

'Risk assessed agent' is defined as that agent who the risk assessing agent considers a possible agent with whom to form a business association in a collaboration based activity. In other words, this agent is potentially capable of giving the risk assessing agent what he desires or what he wants to achieve in the interaction. As a result, this is the agent which is assessed for the possible perceived transactional risk present in interacting with it.

Fig. 3.3 Risk assessed agent

3.2.4 Physical Environment

'Physical environment' is defined as that domain or interaction paradigm where the agents interact with each other, either physically or face-to-face. An example of this is the traditional way in which the interactions are carried out between any two agents.

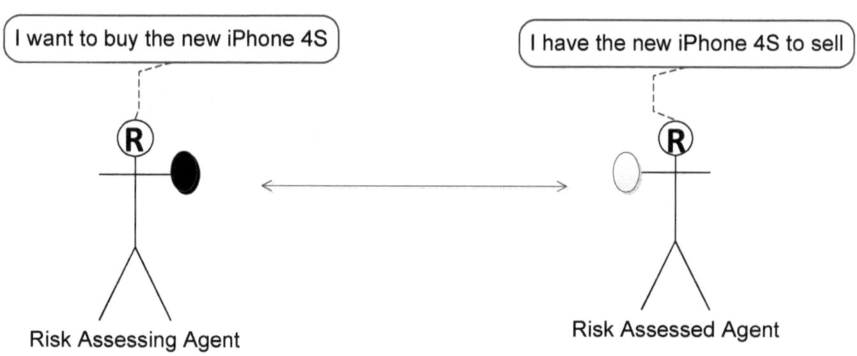

Fig. 3.4 Interaction in a physical environment

3.2.5 Virtual Environment

'Virtual environment' is defined as that domain or interaction paradigm in which the agents interact with each other over a virtual medium. In other words, there is no physical or face-to-face contact between the agents during their interaction.

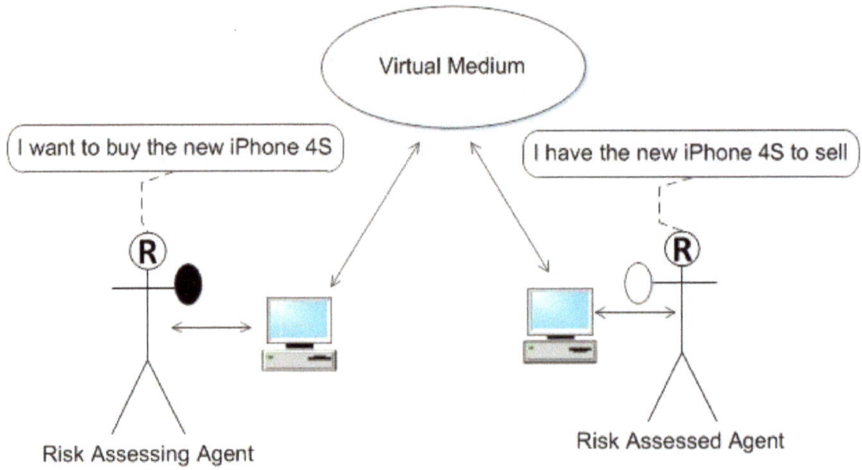

Fig. 3.5 Interaction in a virtual environment

3.2.6 Context

'Context' is defined as the purpose for which an agent carries out a business interaction. In other words, context represents the high level nature of the business activity and gives the description of the risk assessing agent's possible interaction with the risk assessed agent.

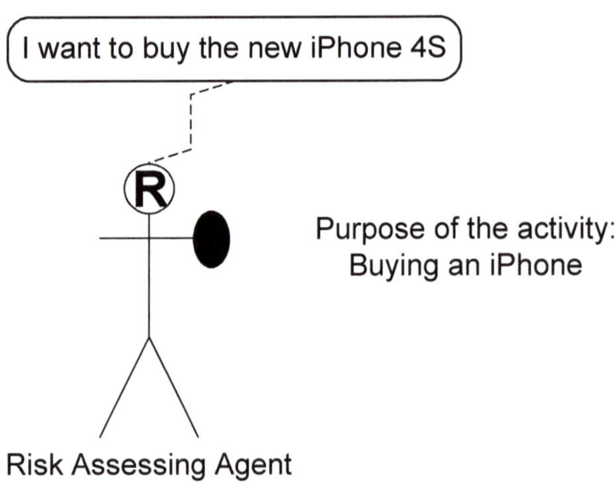

Fig. 3.6 Context of the business activity

3.2.7 Criteria

'Criteria' are defined as the related functionalities or activities that come under a particular context. In other words, the context can be decomposed into several detailed aspects which are regarded as its criteria.

Context: Buying an iPhone		
Criteria:		
Model	iPhone4	iPhone4S
Camera	5 megapixel	8 megapixel
Battery Life	7 hours of talk time	8 hours of talk time
Memory	8 GB	16 GB, 32 GB, 64 GB

Fig. 3.7 Criteria in a business activity

3.2.8 Assessment Criterion and Initial Desired Outcomes

'Assessment criterion' is defined as that criterion from the available criteria which the risk assessing agent wants to achieve in its business interaction. In a business interaction, there may be more than one assessment criterion that is required for its successful completion. The collection of all such assessment criteria is termed the 'initial desired outcomes'. In other words, the initial desired outcomes of the interaction are the set of factors which show specifically what the risk assessing agent wants in his interaction in a particular context.

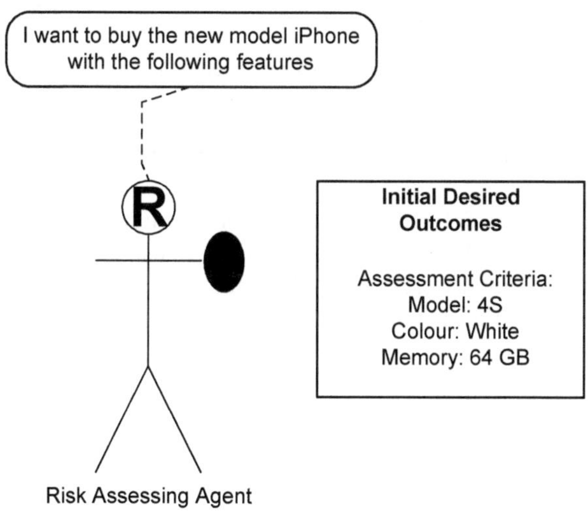

Fig. 3.8 Assessment criterion and initial desired outcomes

3.2.9 Expected Behavior

'Expected behavior' is defined as that behavior which the risk assessing agent expects the risk assessed agent to adhere to in their collaboration-based interaction, before the negotiation phase is carried out. This behavior is formed by the risk assessing agent and is according to the collection of initial desired outcomes and other functionalities, based on how he wants to achieve them while interacting with the risk assessed agent. Further explanation about the formation of expected behavior is discussed in the next chapter.

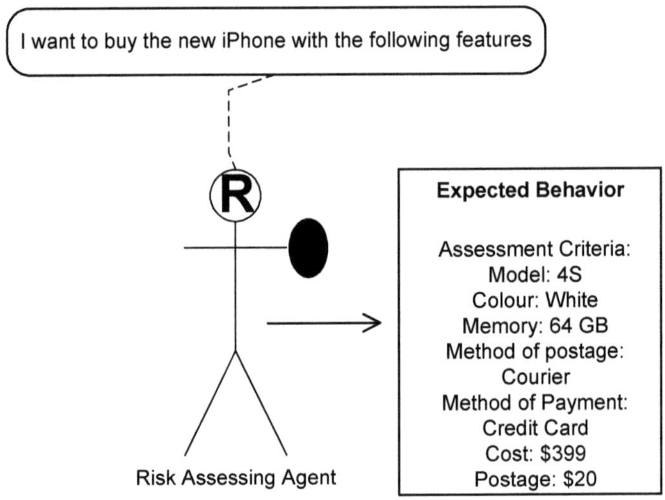

Fig. 3.9 Expected behavior

3.2.10 Expectations or Desired Outcomes

'Expectations' are defined as that behavior which both the agents decide on, after the negotiations. This behavior is an extension of the expected behavior discussed in the last section. When the expectations are formed, then the assessment criteria in it form the desired outcomes that are expected from the business activity.

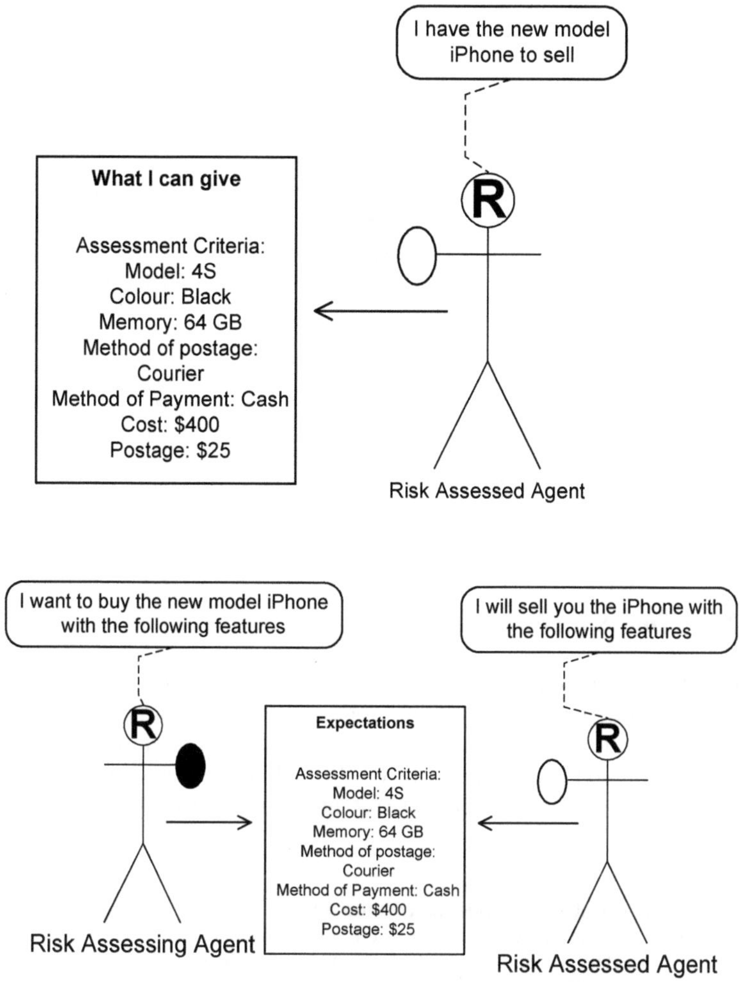

Fig. 3.10 Formation of expectations and desired outcomes

3.2.11 Actual Behavior

'Actual behavior' is defined as that behavior of the risk assessed agent, by which he acts in the interaction with the risk assessing agent. In other words, actual behavior is the quantitatively expressed set of functionalities or activities delivered by the risk assessed agent in his interaction with the risk assessing agent.

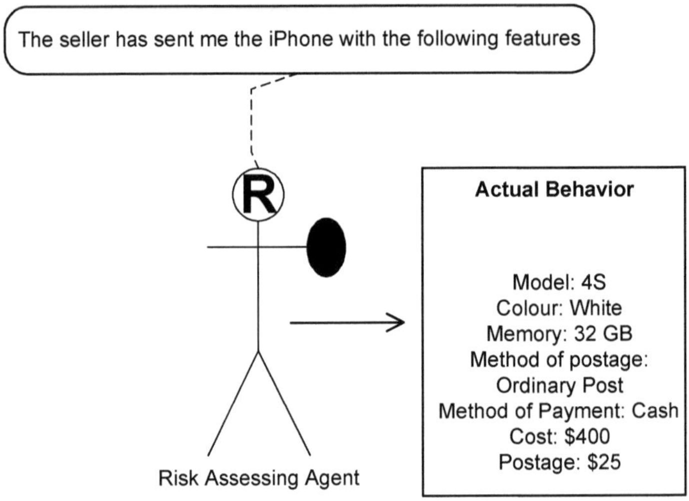

Fig. 3.11 Actual behavior

3.2.12 Time Space

'Time space' is defined as that total period of time which the risk assessing agent takes into consideration to ascertain the perceived transactional risk of forming a business interaction with the risk assessed agent. The duration of the time space varies according to the period of time over which the risk assessing agent wants to determine the possible transactional risk in forming an interaction with the risk assessed agent.

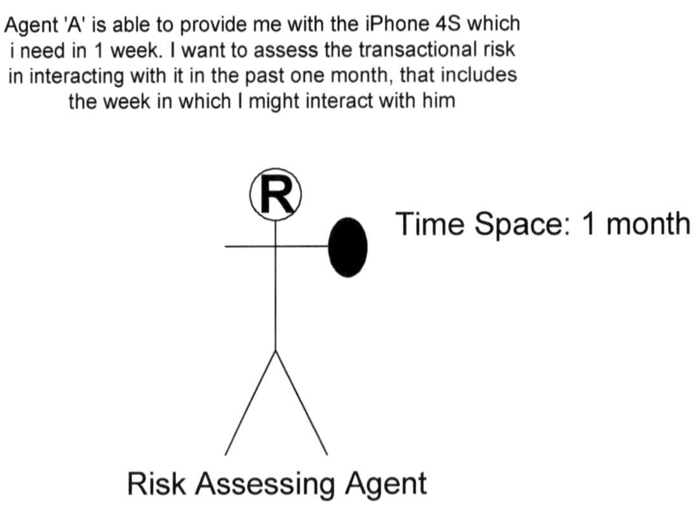

Fig. 3.12 Time space of the business activity

3.2.13 Time Slot

'Time slot' is defined as a non-overlapping interval of time in the time space of the interaction. The time slot is obtained by dividing the time space into different equal non-overlapping parts of time.

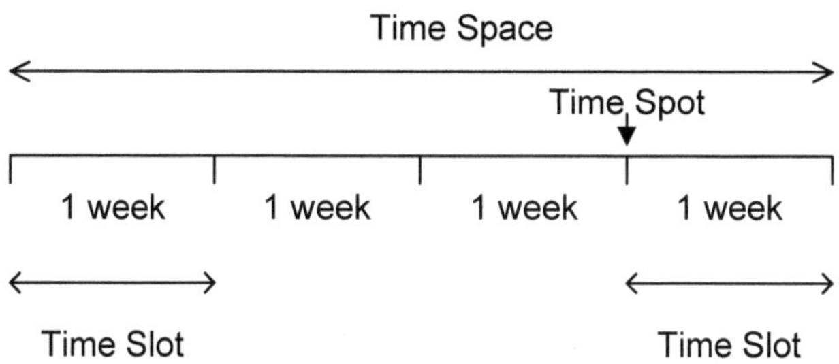

Fig. 3.13 Time slot in the business activity

3.2.14 Time Spot

'Time Spot' is defined as that spot of time in the time space in which the risk assessing agent initiates its business interaction with the risk assessed agent.

3.2.15 Semantics

'Semantics' is defined as the meaning of the concepts and relationships.

3.2.16 Trustworthiness

'Trustworthiness' is defined as the capability of an agent to complete a certain task. The trustworthiness of an agent to complete a task is determined from his own past interaction history with the interaction initiating agent.

3.2.17 Reputation

'Reputation' is defined as the capability or the dependability of an agent to complete a certain task. The reputation of an agent to complete a task is determined from the recommendations of other agents.

3.2.18 Recommending Agent

'Recommending agent' is defined as that agent who gives his/her opinion to the risk assessing agent, when he is solicited in order for them to determine the ability of the risk assessed agent to complete a task.

3.2.19 Resources

'Resources' are defined as that item which the risk assessing agent invests in the interaction to achieve his desired outcomes while interacting with a risk assessed agent. The resources might vary according to the context of the interaction.

Fig. 3.14 Resources in the business activity

3.2.20 Possible Consequences of Failure

'Possible consequences of failure' is defined as the loss which may be experienced by the risk assessing agent as a result of non-achievement of his desired outcomes in the interaction with the risk assessed agent.

Based on the concepts defined, in the next section we present our definition of perceived transactional risk in business activities.

3.3 Definition of Perceived Transactional Risk in Business Activities

We define perceived transactional risk in business domain activities as the *likelihood* that the *risk assessing agent* might not achieve his *desired outcome* from the interaction due to either the *risk assessed agent not acting as expected* according to the *expectations* or the occurrence of uncertain events, in the given *context* and *time* once the interaction begins, resulting in the *loss of investment and resources* involved in the business activity.

This definition comprises many important terms which are pivotal in determining the perceived transactional risk in the business interaction. We have explained some of these terms such as risk assessing agent, risk assessed agent, and context in the previous section. The other terms that are italicized in the definition are further explained in the subsections below.

3.3.1 Likelihood

Likelihood arises from the uncertainty associated in the interaction and refers to a possibility or doubt which is not clearly understood or too readily predicted. This is possible when the risk assessing agent wants a certain outcome in an interaction to happen, but is not sure of what the actual outcome is going to be. This represents a doubt in the risk assessing agent's mind while determining the direction in which the interaction might proceed. In this definition, the term 'likelihood' is associated with the possibility of the interaction proceeding in a negative or unwanted way, rather than in a neutral or a positive way. In reality, a positive or a neutral outcome will not result in the outcome of the interaction being 'unwanted' in the risk assessing agent's point of view as much as a negative outcome will. Hence, while defining perceived transactional risk, likelihood is associated only with the occurrence of negative outcomes.

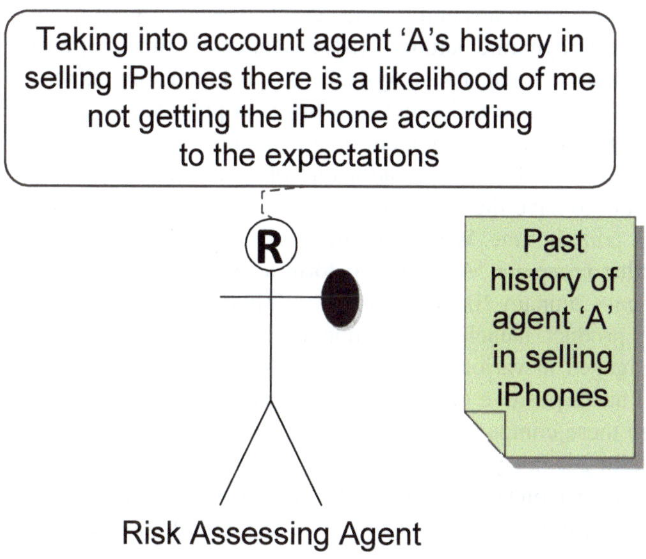

Fig. 3.15 Likelihood of not achieving desired outcomes

3.3.2 Not Acting as Expected

As mentioned earlier, the risk assessing agent defines the initial desired outcomes according to the specific assessment criteria that he wants to achieve in the business activity. He then engages in a negotiation phase with a risk assessed agent to formulate the expectations or desired outcomes of the business association. Further explanation of how the risk assessing agent forms the expectations of his interaction with the risk assessed agent are given later in this chapter.

If the risk assessed agent commits and behaves exactly according to the expectations of the interaction, then he successfully gives the risk assessing agent his desired outcomes from dependent events of the interaction. On the other hand, when these expectations are not adhered to by the risk assessed agent in his actual behavior, then we term this as the risk assessed agent 'not acting as expected'. When the risk assessed agent may not act as expected in an interaction, there is a likelihood that the expectations of the risk assessing agent will not be committed to or met in the interaction by the risk assessed agent.

3.3.3 Time

In Section 2.4.3, we discussed that risk is dynamic and it varies from time to time. The dynamic nature of risk in the described scenario (a person being caught by the police if he drives while drunk at the same time as police are checking) occurs from an event happening during a specific time period. However, apart from this, there is another important dynamic nature of risk dependent on time that needs to be considered during transactional risk analysis. This dynamic nature results from the varying ability of a risk assessed agent or a business provider to deliver the business outcomes to the risk assessing agent on a continuous basis as decided in the expectations. As a result of this dynamic nature of time, it is not possible for the risk assessing agent to have the same impression of the risk assessed agent as he had at a particular point in time. We explain this further in the next paragraph.

Let us consider that company 'A' wants to form a business association with Software Development Company 'B' in order to develop a critical component that is required for the product launch in six months. Also, let us consider that company 'B' has engaged in various projects with 'A' in the past, all of which were successful and had a positive outcome. So, by considering the past business associations between these companies, it can be said that the transactional risk of the current project failing is fairly minimal. However, let us consider that once the project starts, some key personnel working on the project change. The occurrence of such events will present a risk for company A in meeting the project deadline and experience consequences, as shown in Figure 3.16. This risk, which was not there when the business association between them started, arose as a result of several events which occurred after the association started and will have a detrimental effect on the successful completion of the project. Therefore, this dynamic nature of risk, too, is important to consider and appropriate steps for its management should be implemented.

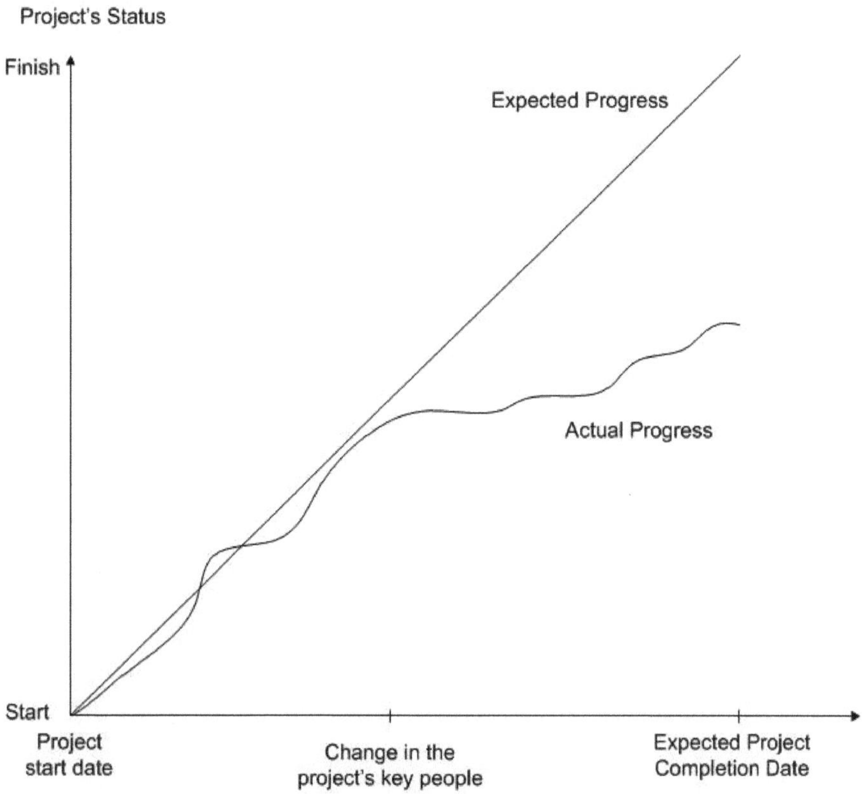

Fig. 3.16 Dynamic nature of transactional risk dependent on time due to change management

Other scenarios where it is important to consider the dynamic nature of risk while making an informed decision may arise from the operational culture of a business provider. As mentioned in Chapter 2, the transactional risk which a risk assessing agent has to analyze could be in a set of business activities with the risk assessed agent that might extend to a point in time in the future. Let us consider that company 'A' forms a business association with Software Development Company 'B' that extends over a one-year period of time in the future. Let us assume at the time of forming a business association, company 'A' has had previous successful business associations with company 'B', but over a period of time, after the current business association starts, the level of company 'B's actual service provided is below expectations. This could be due to various reasons such as complacency on the part of company 'B' as a result of its business flourishing, as shown in Figure 3.17, or a result of varying demands on company 'B's service from other business users, as shown in Figure 3.18.

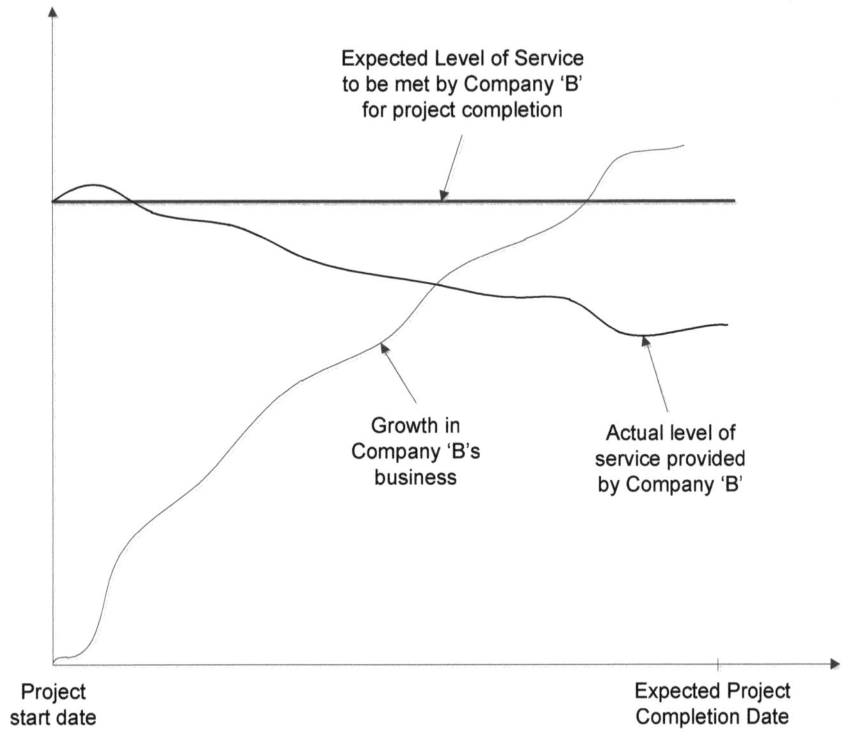

Fig. 3.17 Dynamic nature of transactional risk dependent on the culture of company 'B'

No matter what the reason may be, in each case, there is a risk to company 'A' in meeting the expectations, which was not there at the time of starting the business contract but has crept in due to various factors over a period of time. In each scenario, the inability of the risk assessed agent to act according to the expectations is dynamic and varies from time to time, thereby having a direct impact on the perceived transactional risk in interacting with it. It is important to capture such variation in risk over a period of time during transactional risk analysis in order to make an informed interaction-based decision.

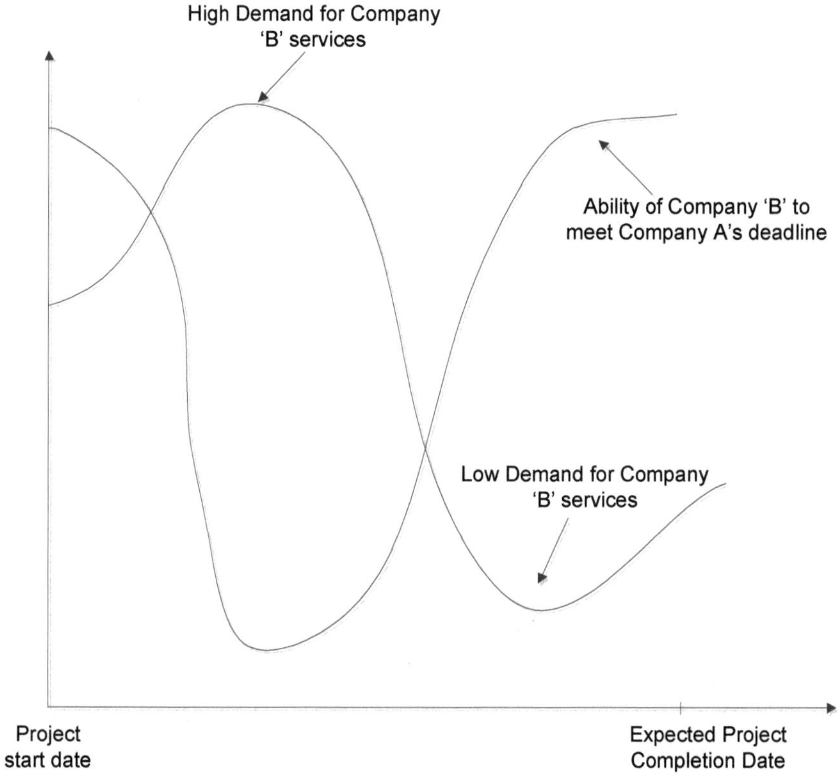

Fig. 3.18 Dynamic nature of transactional risk to Company 'B' dependent on the demands on company 'A's service from other business users

As discussed in Section 1.6, trust in such situations would help company 'A' to make an informed decision in forming a business association with company 'B'. As the outcome of the previous interactions between these agents was positive, the level of trust may naturally be high. However, the events that may lead to the threat of company 'A's project with Company 'B' to fail occur after the business association starts. Even though this may reduce the level of trust which is present, an informed analysis of these events and their impacts is best determined and represented by the analysis of transactional risk over that period of time. Hence, along with trust, it is important to analyze transactional risk in a business activity when making an informed interaction-based decision. This is further important when the risk assessing agent's interaction with the risk assessed agent might start at a point in time in the future or continue till a point in time in the future. Hence, while undertaking transactional risk assessment, it is imperative that the risk assessing agent considers the 'fresh' status of the risk assessed agent and the variability in it, according to the time of his interaction.

In Chapter 4, we explain in detail the concepts that we consider in our approach to consider the dynamic nature of time while ascertaining the perceived transactional risk in the interaction.

3.3.4 Loss of Investment and Resources Involved in the Interaction

As discussed in Chapter 2, while one subcategory of perceived transactional risk signifies the likelihood of the risk assessing agent not achieving his desired outcomes as expected in the interaction, the other subcategory specifies the loss that the risk assessing agent might incur as a result of this. Such possible loss will be in the resources in which the risk assessing agent has invested and has at stake, in exchange for achieving his desired outcomes. These resources vary according to the domain in which the interaction is taking place. In the domain of business activities, there may be various possible scenarios, but in the context of the discussion of this book, we consider that the risk assessing agent interacts with a risk assessed agent to achieve his desired outcomes in exchange for monetary financial value, and hence, the possible loss that he could incur in the interaction is in the financial resources which he has invested and has at stake in the interaction. By 'loss of investment' and 'resources', we mean the financial loss that could be experienced by the risk assessing agent in the resources which he has invested and has at stake while interacting with the risk assessed agent.

The definition of perceived transactional risk encompasses its subcategories and its characteristics in the domain of business interactions. In the next section, we present our approach by which the risk assessing agent can analyze the perceived transactional risk in the domain of business activities.

3.4 Approach for Perceived Transactional Risk Consideration and Analysis in Business Activities

The approach that we utilize for perceived transactional risk consideration and analysis is a 5-step process that consists of *risk identification, risk assessment, risk evaluation, risk management and risk recording.* Figure 3.19 represents the flow of control between these steps. We give a brief explanation of each step below.

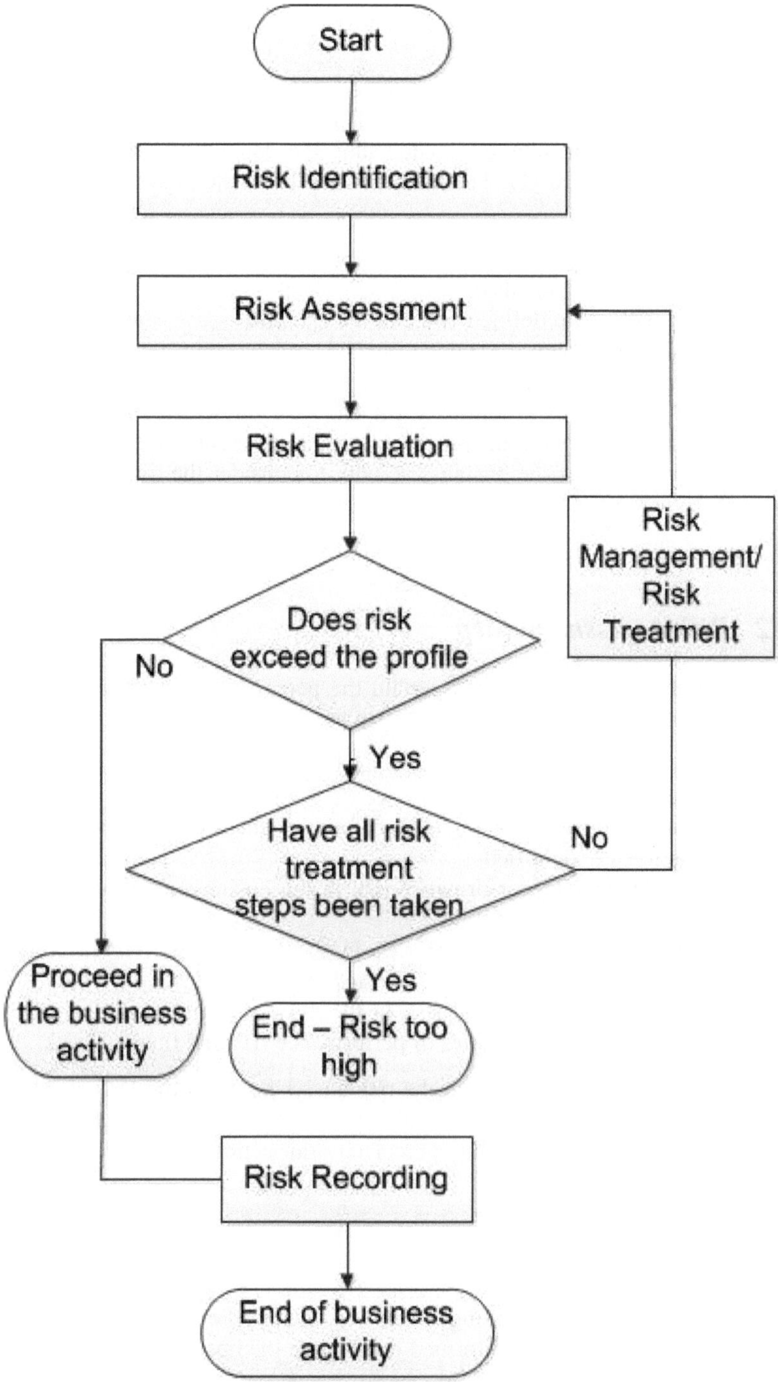

Fig. 3.19 Flow of control between the different steps of perceived transactional risk analysis

3.4.1 Risk Identification Step

This step deals in the formalization of the context, scope and expectations of the business activity for risk assessment. This steps maps with the steps of communication and consultation, establishing the context and risk identification steps of the ISO standard [1]. The main tasks to be carried out in this step are as follows:

1. the context of the business activity for transactional risk analysis is identified and established.
2. further within the defined context, the specific assessment criteria that are desired to be achieved are determined.
3. negotiations are carried out to form the expectations of the business activity. These expectations form the base on which the risk assessing agent has to carry out the process of risk assessment in the next steps.
4. the time space of the business activity to consider the time specific nature of the perceived transactional risk is formulated.

The steps involved in this process are explained in detail in Chapter 4.

3.4.2 Risk Assessment Step

This is a multi-stage process to ascertain the perceived transactional risk in the business activity. The main tasks in this step are:

1. the risk assessment process to analyze the level of transactional risk is initiated.
2. the various different types of uncertainties that could be possible in the interaction are identified.
3. the subcategory of performance risk in the business activity according to the expectations is determined.
4. the subcategory of financial risk in the business activity according to variations in investment in the interaction is determined.
5. the perceived transactional risk in the business activity is determined by combining the subcategories of performance risk and financial risk.

The steps in this process map to the risk analysis step of the ISO standard [1]. In Chapters 5-7, we explain in detail our proposed approach by which the risk assessing agent ascertains the perceived transactional risk in his business activity.

3.4.3 Risk Evaluation Step

In this step, the risk assessing agent evaluates the ascertained perceived transactional risk to make an informed decision in the business activity. The main tasks in this step are as follows:

1. the risk tolerance level or risk attitude of the risk assessing agent in the business activity is determined.
2. the determined level/s of perceived transactional risk is evaluated against the risk attitude or risk profile of the risk assessing agent.
3. based on the analysis results, an informed interaction-based decision about the future course of the interaction is undertaken. This analysis forms input to the process of risk management.

This step maps to the risk evaluation step of the ISO standard [1]. In Chapter 8, we explain in detail the proposed approach by which the risk assessing agent can evaluate the perceived transactional risk in interacting with a risk assessed agent and utilize it to make an informed interaction-based decision.

3.4.4 Risk Management Step

In this step, the risk assessing agent takes appropriate steps to ensure the risks are addressed and managed so that the outcome of the interaction is either positive or neutral. This is an iterative step that may involve more than one action to be carried out more than one time with the aim of minimizing the transactional risk to an acceptable level. The main tasks in this step are as follows:

1. formalize appropriate steps or plans by which the identified risk events can be addressed.
2. implement the plans to mitigate the risks.
3. regularly assess and evaluate the risks to determine whether they have been minimized to an acceptable level. This analysis forms input into the decision-making process.

This step maps to the risk treatment step of the ISO standard [1]. It is important to note that the process of risk management is interaction- specific and it is not possible to define a framework that can be generalized for every business activity. In Chapter 9, we discuss a business activity scenario of just-in-time production and explain how a manufacturer can treat and manage the identified risks of not meeting the customers' demands.

3.4.5 Risk Recording Step

This step records the actual level of transactional risk present in the business association, if the risk assessing agent proceeds in forming a business interaction with the risk assessed agent. The recorded level of risk will be utilized in either risk education or recommendation communication. It is important to note that the determined level of transactional risk excludes that which can be experienced from non-dependent events. The main tasks in this step are as follows:

1. capture the actual behavior of the risk assessed agent in relation to the expectations of the interaction. Furthermore, determine the accuracy with which the expectations were communicated to the risk assessed agent and the significance of each assessment criterion.

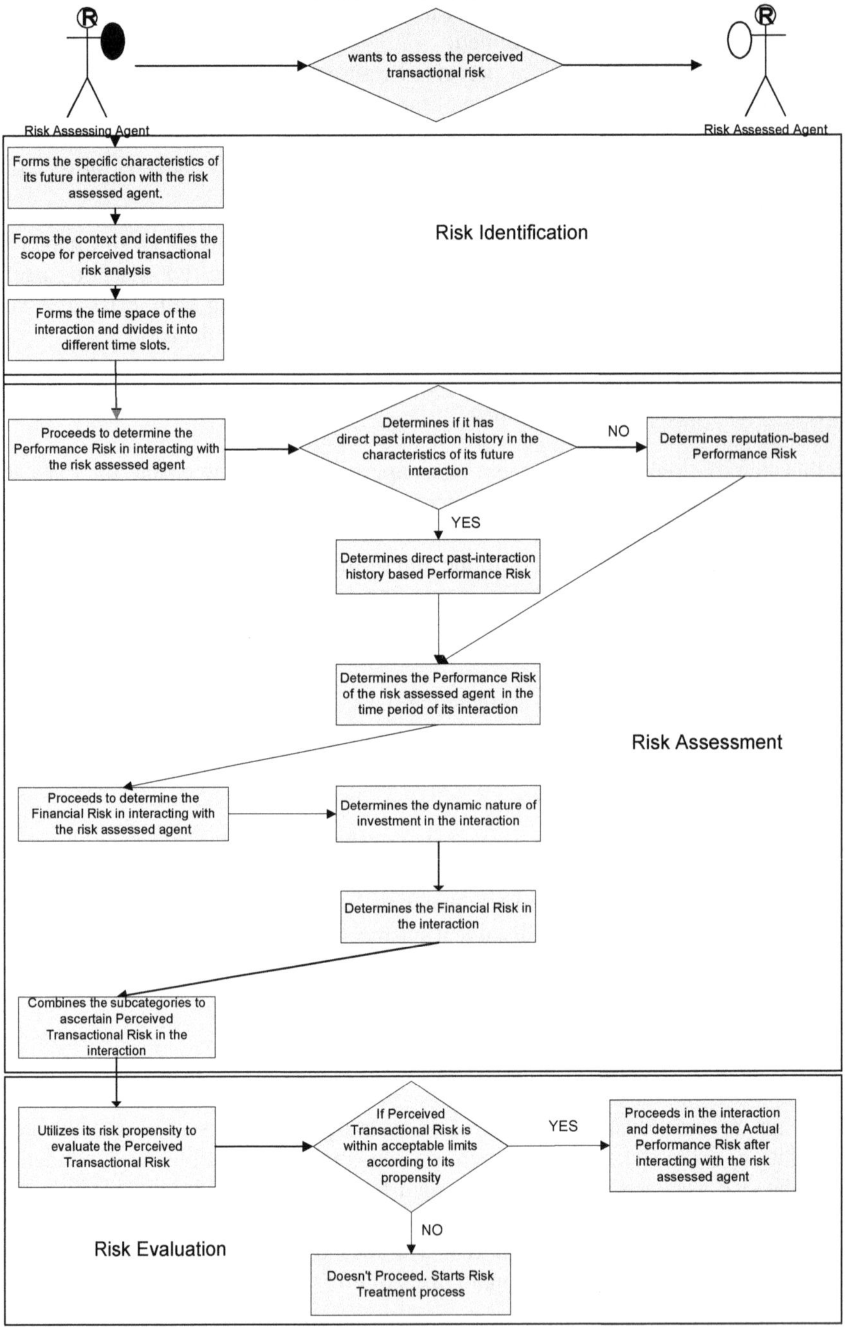

Fig. 3.20 Steps in perceived transactional risk analysis in business activities

2. quantify the expected behavior of the risk assessed agent from the expectations of the interaction.
3. ascertain the level of non-commitment in the interaction.
4. develop the 'risk set' that formalizes the summary of the risk assessing agent's interaction with the risk assessed agent.

This step maps with the risk recording step of the ISO standard [1]. In Chapter 9, we explain in detail the proposed approach by which the risk assessing agent can evaluate the actual level of perceived transactional risk present in his interaction with the risk assessed agent and formulate the 'risk set' of the business activity.

Each step of transactional risk analysis discussed above contains various sub-steps. Figure 3.20 details the steps of risk identification, risk assessment and risk evaluation. It should be noted that the relationships shown between the different entities in Figure 3.20 represent the flow of control in the proposed approach, and do not represent the relationships in the object modeling language.

In the next section, we discuss two different ways to initiate the process of risk assessment in business activities.

3.5 Ways to Initiate the Process of Risk Assessment

There are two broad ways by which a risk assessing agent can initiate the process of risk assessment. They are:

1. using own past experiences
2. soliciting recommendations

3.5.1 Using Own Experiences

Own or self-experience is the direct first-hand experience of the risk assessing agent with the risk assessed agent in question. Such personal experience may be from a part or the whole of a current association or collaboration between the users. For example, risk assessing agent 'A' interacted with risk assessed agent 'B' last week to buy an iPhone 4S. This previous interaction experience of agent 'A' with agent 'B' can be utilized by agent 'A' in his next interaction to initiate the process of risk assessment, if it is in the same context and appropriate time period as shown in Figure 3.21.

But in a real life scenario, finding such own experience between any two agents that are in the same context, assessment criteria and time period to consider the context-specific nature of transactional risk is not always possible. In such cases, recommendations from other agents are utilized to initiate the process of risk assessment.

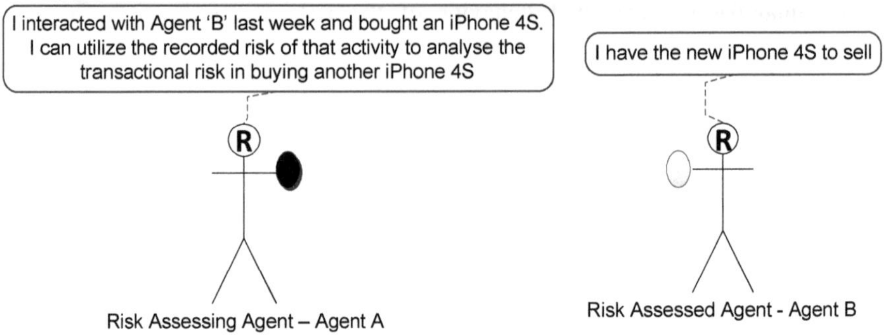

Fig. 3.21 Using own experience to initiating the process of risk assessment

3.5.2 Soliciting Recommendations

In the absence of own experiences, the process of risk assessment is begun by a third-party agent who provides an introduction or recommendation to the risk assessing agent about the risk assessed agent/s capability to deliver the outcome in the solicited criteria. For example, risk assessing agent 'A' wants to buy an iPhone 4S from agent 'B', but does not have any knowledge of his capability in this context. In such a case, as shown in Figure 3.22, he asks for recommendations from other agents to initiate the process of risk assessment in forming a collaboration with agent 'B'. It could also be the case that agent 'A' has his own past experience with agent 'B', but not in the specific context, as required in this business scenario. As discussed earlier, in such a case, agent 'A' cannot use past interaction history and has to rely on a recommendation-based introduction from other agents to initiate the process of risk assessment. Another important factor when considering recommendations from other users is their credibility while communicating those. We will discuss this factor further in Chapter 5.

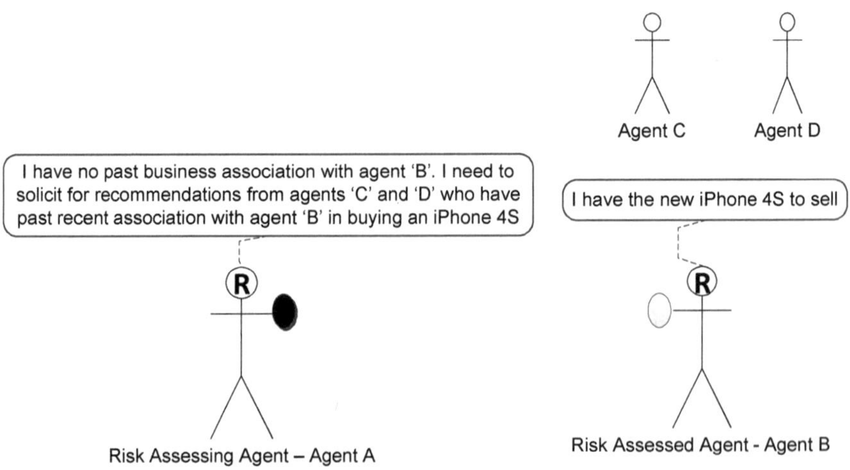

Fig. 3.22 Soliciting recommendations to initiate the process of risk assessment

In both these cases, to initiate the process of risk assessment, the risk assessing agent utilizes the *risk relationship* that he has from either his own past experience or from the recommendations communicated by other agents. In the next section, we will give a detailed explanation of the risk relationship formed between any two interacting agents.

3.6 Risk Relationship Formed between Interacting Agents

We consider that whenever two agents interact with each other, they form a relationship. 'Risk relationship' between two agents relates to the elements of perceived transactional risk which form as a result of their interaction or collaboration. This relationship details the various elements of transactional risk and is conditioned by the specific characteristics of the business activity; these are context specific, assessment criteria specific and dynamic properties. Furthermore, such a risk relationship formed between any two agents not only represents the transactional risk that was perceived before initiating the business association but also represents the actual level of transactional risk that was present in the business association. This actual level of transactional risk is used in the risk recording step (Section 3.4.5) of perceived transactional risk analysis. As shown in Figure 3.23, the recorded level of transactional risk will then be used when the risk assessing agent either wants to:

(a) utilize his past business association with the risk assessed agent in analyzing the level of transactional risk in his future interaction, or
(b) when solicited, communicate recommendations to another risk assessing agent about the risk assessed agent.

3.6.1 Dependent Factors of a Risk Relationship

It is important to note that a risk relationship exists between any two agents only if they interact with each other. So, as shown in Figure 3.24, the relationship between the risk assessing agent and the risk assessed agent is represented by a ternary association relation between them. During the risk assessment step, if a risk assessing agent assesses the perceived transactional risk in forming an interaction with a risk assessed agent, and later decides against forming a collaboration with that agent for that activity, then no risk relationship is formed between these two agents for that particular interaction. So, the risk relationship between any two agents is realized by an interaction between them. The risk relationship, in turn, is dependent on the different factors of the sub-categories of transactional risk related to the interaction. These are the probability of failure of

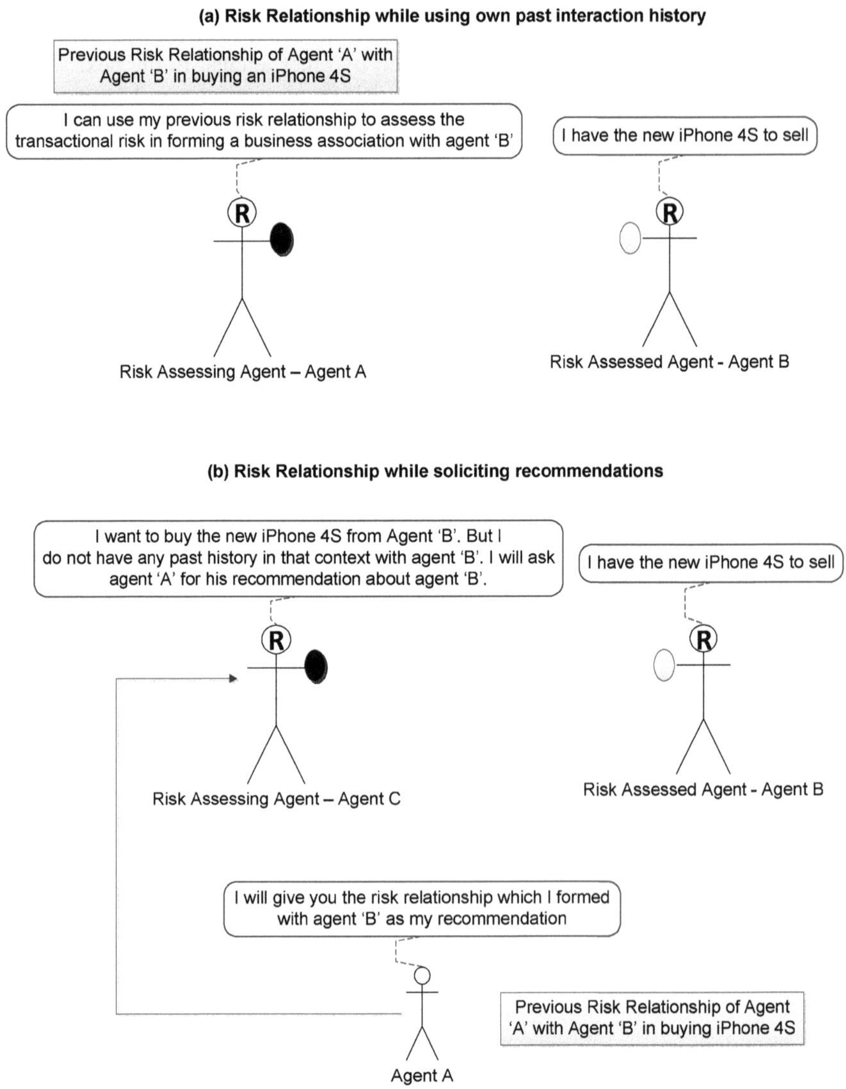

Fig. 3.23 Importance of a risk relationship while conducting risk assessment

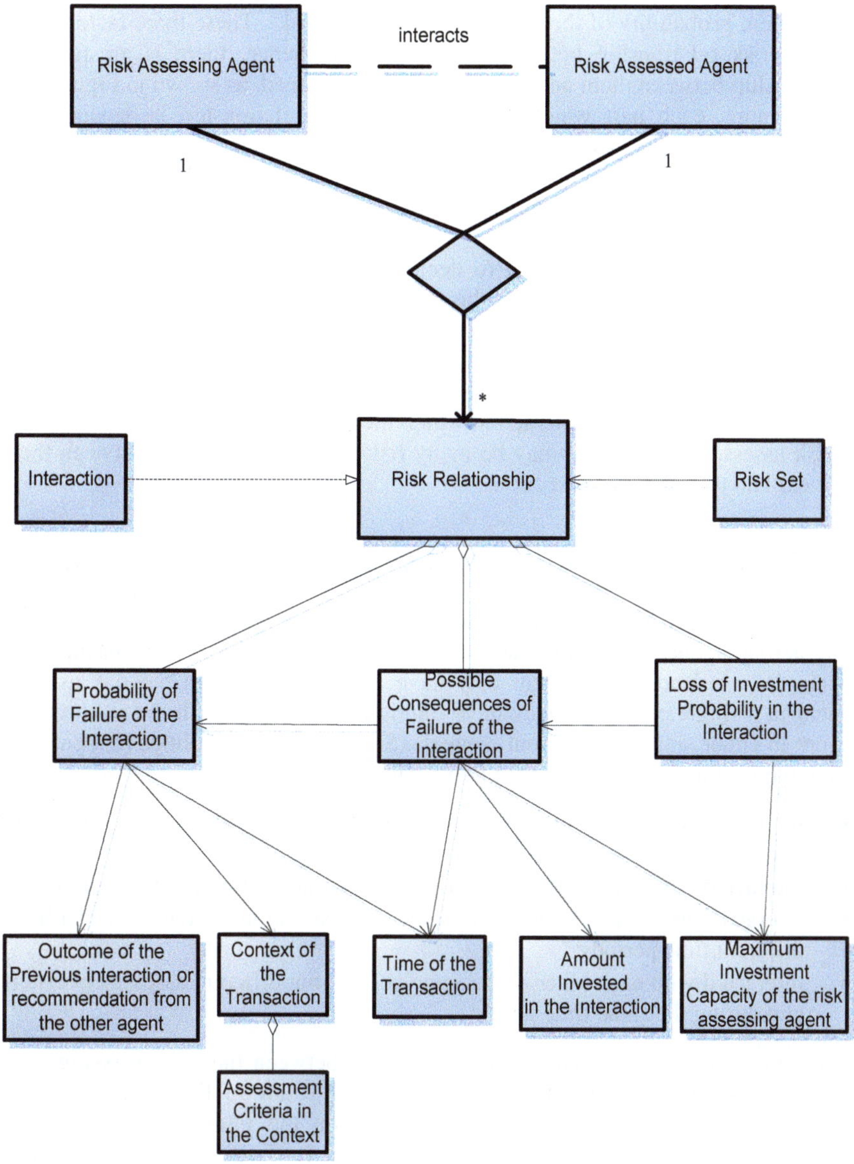

Fig. 3.24 Risk relationship formed between two interacting agents

the interaction (performance risk), possible consequences of failure and the loss of investment probability of the interaction (financial risk). These three factors make up the risk relationship between the agents and hence, there is an aggregate relationship between them and a risk relationship formed, as shown in Figure 3.24. Furthermore, each part which makes up the risk relationship is dependent on different sub-factors, as shown in Figure 3.24. Each individual sub-factor is utilized to determine the part of the risk relationship which is dependent on them, and hence, there is a dependency relationship between the factor and the individual sub-factors.

So, for a risk assessing agent to decide whether to consider a particular risk relationship, it needs to study the specific characteristics from which the relationship was formed. Without these characteristics, a risk relationship does not have any meaning, and without a risk relationship, past interaction history does not have any meaning to the risk assessing agent during transactional risk analysis. Depending on the type of interaction which the risk assessing agent may have with a risk assessed agent, there may be many risk relationship formed between them. We explain this in the next section.

3.6.2 Number of Risk Relationships between Any Two Agents

There are many different possible interaction scenarios between any two interacting agents that will lead to different types of risk relationships. For example, a risk assessing agent may have an interaction with the risk assessed agent in a single context or may have multiple interactions with the risk assessed agent in either single or different contexts. Some of the possibilities to mention in such types of activities are that the two agents might have:

- interacted with each other in different contexts but during the same time period;
- interacted with each other in different contexts and different time periods;
- interacted with each other in the same context and assessment criteria but at different time periods;
- interacted with each other in the same context but different assessment criteria and during the same time period.

No matter what type of interaction is involved between the risk assessing agent and the risk assessed agent, each activity will result in the formation of an individual risk relationship. So, depending upon the interaction history, it is possible that the risk assessing agent may have a 1…1 or 1...m cardinality in its risk relationship with a risk assessed agent, as shown in Figure 3.25.

However, as discussed in the earlier section, each risk relationship is dependent on the specific subcategories and characteristics of perceived transactional risk of their subsequent interaction. So, for a risk assessing agent to decide whether to consider a particular risk relationship, it needs to study the specific characteristics from which the relationship was formed.

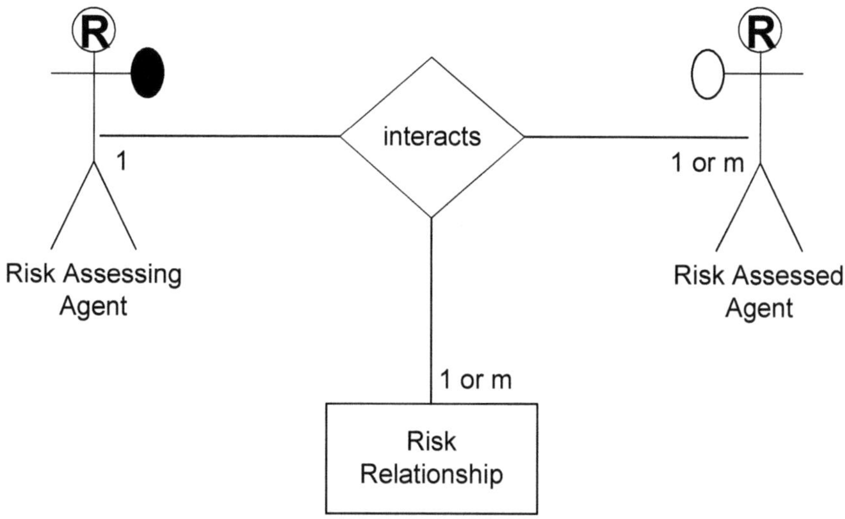

Fig. 3.25 Different interaction scenarios and the risk relationships formed

3.6.3 Risk Set for Representing the Risk Relationship

As mentioned earlier, if the risk assessing agent has its own past interaction history, then it utilizes the risk relationship from its previous interaction experience with the risk assessed agent to initiate the process of risk assessment. In the case where there has been no direct or own past interaction history between them, the risk assessing agent, in initiating the process of risk assessment, utilizes recommendations from other agents, by issuing a reputation query to solicit recommendations from other agents. In this query, the risk assessing agent specifies the context, assessment criteria and timeframe in which it wants the recommendations from the other agents in relation to a risk assessed agent. The agents who have a past interaction history with the risk assessed agent in question and in the specific characteristics, reply with their recommendations to the risk assessed agent.

However, in order for a risk assessing agent to properly assess the risk relationship either from itself or from the recommending users it is important that the recommendations are expressed in a standard and unified way. To facilitate this, we propose the '*risk set*'. The risk set is a standard format for either the risk assessing agent to consider its own historical data or for the recommending agent to give its recommendations about a risk assessed agent in a standardized and efficient way. The risk set is dependent on the risk relationship and hence as shown in Figure 3.24, there is a dependency relationship between them. The format of the risk set is defined below:

{RA1, RA2, Context, AFL, (Assessment Criterion, Commitment level), Cost, Start time, End time}

where:

RA1 is the risk assessing agent in the interaction, which is also the recommending agent while giving recommendations.

RA2 is the risk assessed agent in the interaction.

Context represents the context of the interaction.

AFL represents the 'Actual FailureLevel' of the risk assessed agent determined after the business activity by assessing the level of non-commitment in his actual behavior with respect to its expectations.

(Assessment criterion, Commitment level) 'Assessment criterion' represents the assessment criterion in the expectations of the interaction with the risk assessed agent. The combination of (Assessment criterion, Commitment level) is represented for each assessment criterion in the expectations. This is the set of factors on which the risk assessing agent or the recommending agent interacted with the risk assessed agent and later assigned it the 'Actual FailureLevel' (AFL) in its interaction. It is important to mention these critera while giving recommendations so that a risk assessing agent who solicits recommendations knows the assessment criteria on which this particular risk assessed agent has been assigned the FailureLevel (AFL) value. In this way, the risk assessing agent can consider only those recommendations which are of interest to him, according to the expectations of his future interaction. 'Commitment level' specifies whether or not the particular assessment criterion was fulfilled by the risk assessed agent, according to the expectations of the interaction. A value of either 0 or 1 is assigned to it, based on the commitment of the risk assessed agent for that criterion. A value of 0 signifies that the assessment criterion was not fulfilled by the risk assessed agent, according to the expectations, whereas a value of 1 signifies that the assessment criterion was fulfilled, according to expectations. Further explanation is given in Table 3.1.

Cost represents the total financial value at stake in the interaction.

Start Time is the time at which the interaction with the risk assessed agent started.

End time is the time at which the interaction with the risk assessed agent ended.

Table 3.1 The commitment level of each assessment criterion

Commitment Level	Semantics of the Value
0	The risk assessed agent did not commit to the assessment criterion as expected, according to the expectations
1	The risk assessed agent committed to the assessment criterion exactly according to the expectations

Using the risk set, the risk assessing agent can comprehend the past historical data either from its own history or from other agents in an efficient way. This will be beneficial for it while ascertaining the performance risk in its activity, which we will discuss in the next chapter.

3.7 Recapitulation

In this chapter, we presented the key concepts that we will utilize in this book for risk analysis. We then presented our definition of perceived transactional risk that takes into consideration its subcategories and characteristics according to its object of analysis in business activities and defined our framework by which the risk assessing agent can carry out the steps for risk analysis in his business activity with the risk assessed agent. A brief overview of the different steps in our proposed framework for risk analysis was discussed, followed by the different approaches by which the risk assessing agent initiates the process of risk assessment. We then discussed the risk relationship and risk set that is formed as a result of collaboration between any two users. In the next chapter, we present details on how the risk assessing agent initiates the risk assessment process and determines the performance risk in his business activity with the risk assessed agent.

Reference

[1] Standards Australia: AS/NZS ISO 31000:2009 Risk Management - Principles and guidelines. In: Standards Australia (ed.), pp. 1–29 (2009)

Chapter 4
Risk Identification in Business Activities

4.1 Introduction

In the previous chapter, we discussed the various steps in the process of transactional risk analysis in business activities. In this chapter, we define the main steps that need to be carried out during the process of risk identification. As defined earlier, risk identification deals in formalizing the purpose and scope for which the business activity is being carried out and then identifies those events that will lead to the occurrence of a negative outcome in the business activity. However, before such events can be identified, there are various detailed steps that need to be carried out in order to capture the scope of the business activity. They are as follows:

(a) formalizing the assessment criteria of the business activity,
(b) formalizing the significance of the assessment criteria, and
(c) division of the time to capture the dynamic nature of transactional risk.

We explain each of these steps in the rest of this chapter.

4.2 Formalizing the Assessment Criteria of the Business Activity

In Section 3.2, we mentioned that 'context' of the business activity represents the broad purpose for which the business association is being carried out. However, simply capturing the context does not represent all the information required to carry out the process of risk identification. For example, if we consider the activity 'risk assessing agent 'A' wants to buy an iPhone', even though it represents the context and the broad reason for agent 'A's business association, it does not represent the required minute details in that context, for example, the model of iPhone required (3 or 3GS or 4 or 4S), the amount of memory required (8 GB or 16GB or 32 GB or 64GB), colour required (white or black) type (new or used) etc. All these factors that represent the fine-level of details are termed the *'criteria'* in that context. However, identifying the criteria in a context does not mean that agent 'A' wants to achieve all of the listed criteria to a complete stage.

For example, it may be the case that agent 'A' is a businessman who will be away from his office on business often. During this time, he regularly wants to

O.K. Hussain et al.: Risk Assessment & Management in the Networked Economy, SCI 412, pp. 85–105.
springerlink.com

check his emails and therefore, would like a phone that supports wireless connectivity to the internet. For him, having an iPhone 3 with wifi capability is not enough, as the places where he is usually offsite may not have wifi networks. So agent ‘A’ wants to buy an iPhone 4S that has 3G capability and SIRI functionality to automate his required applications. Hence, in this case, colour and memory capacity do not matter much to agent ‘A’ as does the network connecting capability and functionalities of the phone. As mentioned in the last chapter, such desired outcomes in business criteria are termed the *assessment criteria.* It is important to identify such assessment criteria for formalizing the scope, according to which transactional risk should be determined, as shown in Figure 4.1.

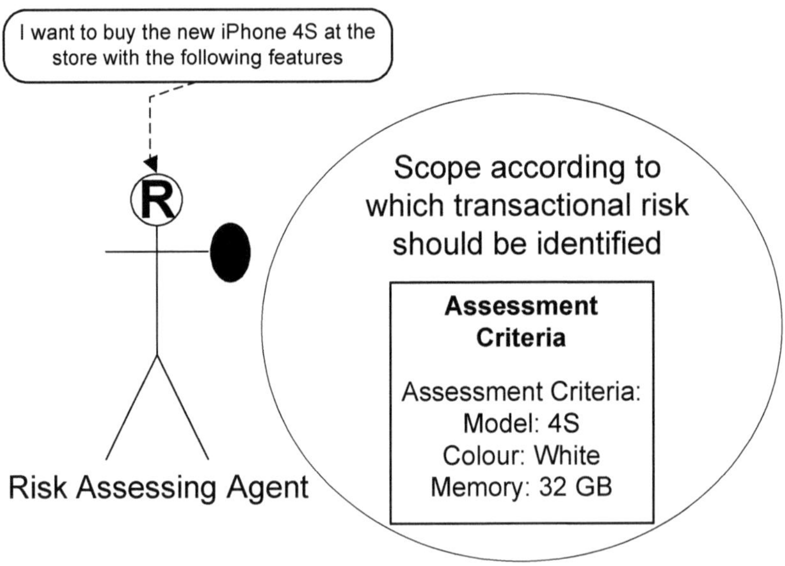

Fig. 4.1 Scope according to which transactional risk should be analyzed

In the context of a business association, there may be more than one assessment criteria which the risk assessing agent may want to achieve as an outcome of his business association, and it is important that all those criteria are formalized to correctly identify the risk in achieving them. However, the process of formalizing the assessment criteria of the business activity depends on the type of business activity. We explain this further in the remaining subsections below.

4.2.1 Formalizing the Assessment Criteria in Single User Activity

In Section 2.5, we mentioned that Single User Activities are those that are dependent only on the risk assessing agent in achieving a successful outcome. Examples of such activities may be the face-to-face activities of the risk assessing agent that will have an instant outcome, as shown in Figure 4.2. In such activities,

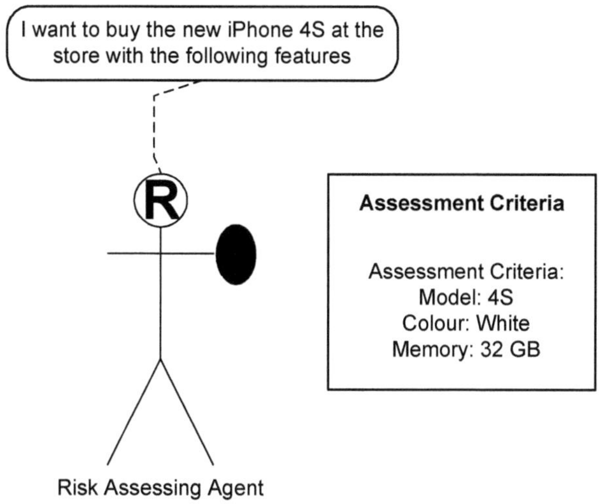

Fig. 4.2 Formalizing the assessment criteria in single user activity

the risk assessing agent is solely responsible for identifying and formalizing the assessment criteria of his business activity.

4.2.2 Formalizing the Assessment Criteria in Collaboration-Based User Activity

The other type of business activity is collaboration-based between the risk assessing agent and one or more risk assessed agent. In such activities, even though the risk assessing agent is responsible for specifying the initial desired outcomes, it may be possible that the business contract or expectations are formed with slightly modified criteria. This may be due to reasons such as the inability of the risk assessed agent to commit to the assessment criteria as desired by the risk assessing agent etc. However, identifying all such scenarios and formalizing the assessment criteria on which both agents agree beforehand is important in order to have a clear and unambiguous representation of the scope on which transactional risk is identified and assessed.

For example, let us consider that agent 'A' wants to buy an iphone 4S from agent 'B'. The initial desired criteria which agent 'A' forms are colour: white and Memory 64 GB. Let us consider that the stock of iPhones which agent 'B has, meets the colour requirement of agent 'A' but are underpowered in terms of memory (having only 32GB). In such a case, if agent 'A' does not communicate the specific details of the assessment criteria to agent 'B' and forms a business association with agent 'B' only on the context of 'buying an iPhone', he will experience a negative outcome in the business activity. This may result in agent 'A' giving a high risk rating to agent 'B' during the process of risk recording, which may affect agent 'B's rating and the chances of him acquiring any future

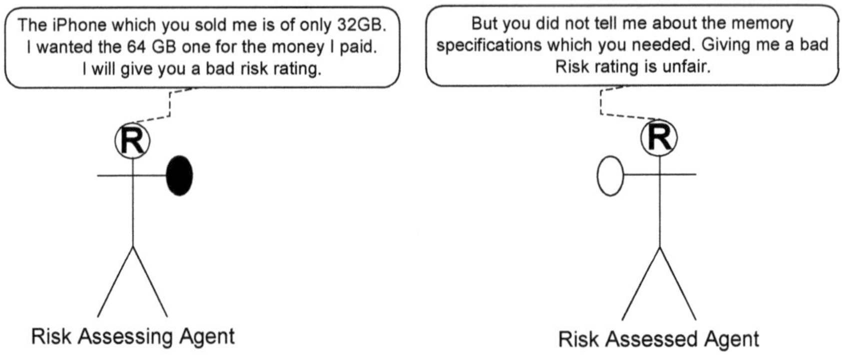

Fig. 4.3 Misunderstandings while formalizing the assessment criteria in a collaboration-based activity

business. Also, scoring a high risk rating in such a scenario is unfair to agent 'B' as the required expectations were not communicated clearly in the context of the business scenario, as shown in Figure 4.3.

4.2.2.1 Negotiation-Based Approach to Formalize the Assessment Criteria in a Collaboration-Based User Activity

So, to formalize and decide on the expectations in a collaboration-based activity, we propose that once the risk assessing agent forms his initial desired outcomes, he then engages in a two-way negotiation phase with a risk assessed agent [1]. The aim of this phase is for the risk assessing agent to first express his initial desired outcomes to the risk assessed agent in clear and unambiguous terminology, in order to determine the risk assessed agent's willingness to commit to them. The risk assessed agent may either accept these or propose a modified set of functionalities in the assessment criteria to which he can commit to. During this process, both the agents may arrive at a set of functionalities which the risk assessed agent promises to commit to, in order to give the risk assessing agent what he desires. The output of the negotiation phase is a set of quantitatively expressed activities and functionalities that the risk assessed agent is capable of, and willing to perform in the interaction. If the risk assessing agent agrees with such a set of functionalities and decides to interact with the risk assessed agent, then he forms the 'expectations' of the interaction.

The formed 'expectations' represent the scope and the way both the agents should act in order to commit to the desired outcomes of the activity. In a business interaction, this is the set of factors against which the risk assessing agent signs a business contract or agreement with a risk assessed agent in the particular context, and subsequently, it is against these factors that the risk assessing agent should identify the transactional risk in the successful completion of the business activity. It is important to note that the risk assessing agent and the risk assessed agent may have to undergo a number of negotiation cycles between them in order to agree and decide on the expectations. Figure 4.4 shows the process by which the risk assessing agent forms the expectations of the interaction in a collaboration-based activity.

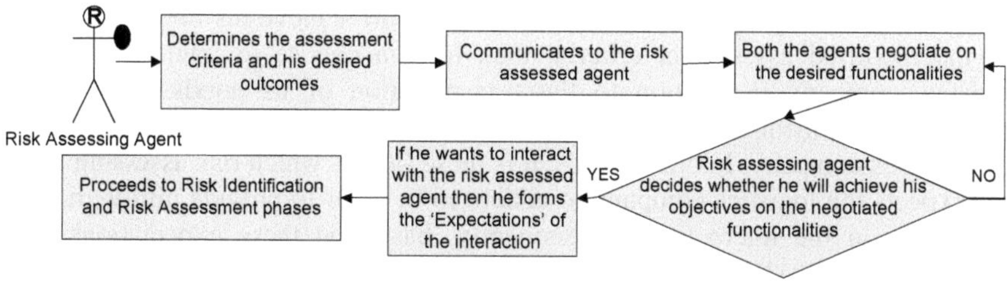

Fig. 4.4 Process by which the risk assessing agent forms the expectations of its business activity

Let us consider the following example: Risk assessing agent 'A' wants to interact with a logistics company in order to transfer his goods from Milan, Italy to Perth, Australia. The context of the interaction is 'Transporting Goods'. Further in the book, we term the context of the interaction as 'C' and the assessment criteria which the risk assessing agent wants to achieve in the particular context as C1, C2, C3……Cn. Let us assume that the desired outcomes and the assessment criteria of the risk assessing agent in his interaction are:

- the goods should reach their destination undamaged (C1)
- the goods should be delivered to the destination by 8 am on the specified day (C2)
- the logistics company should unpack the goods at the destination (C3)
- the logistics company should provide a track and trace facility (C4)
- the logistics company should complete the transaction for 10,000 euros (C5)
- Agent 'A' will pay the logistics company 7,000 euros when the goods are picked up and 3,000 euros in Australian dollars when the goods are delivered to Perth (C6).

Based on the assessment criteria, the risk assessing agent enters into a negotiation phase with the logistics company. In the negotiation phase, the logistics company identifies those factors communicated by the risk assessing agent, which prevent it from fulfilling all the outcomes and meeting the assessment criteria, and subsequently, modifies it according to how it promises to carry it out and then communicates this back to the risk assessing agent 'A'. Let us consider that the assessment criteria modified by the logistics company are:

- the goods should reach their destination undamaged (C1)
- the goods should be delivered to the destination by 12 pm on the specified day (C2)
- the logistics company should unpack the goods at the destination (C3)
- the logistics company should provide a track and trace facility (C4)
- the logistics company should complete the transaction for 10,000 euros (C5)
- Agent 'A' will pay the logistics company 7,000 euros when the goods are picked up and 3,000 euros in Australian dollars when the goods are delivered to Perth (C6)

If the risk assessing agent 'A' concludes that it will achieve his objectives based on the modified assessment criteria, then he can choose to interact with this logistics company in relation to the transportation of its goods. This set of modified assessment criteria will form the 'expectations' of his interaction with the logistics company. In other words, it is these factors which risk assessing agent 'A' expects the logistics company to adhere to in order to provide him with what he desires in the interaction. Subsequently, based on these expectations, risk assessing agent 'A' should identify the risks present and utilize these during transactional risk assessment. The expectations of a business activity identify the scope in which perceived risk should be analyzed. This will also help in establishing the internal context and environment that forms the basis for the risk assessing agent to identify the risks in the successful completion of his business activity. Once the expectations have been formed, then the risk assessing agent should identify the significance of each assessment criteria. We will discuss these in the next section.

4.3 Significance of the Assessment Criteria

In both types of business activities discussed above, it is quite possible that the risk assessing agent may want more than one assessment criterion to be achieved. Further, from the perspective of the risk assessing agent, the assessment criteria in his expectations may not necessarily be of equal importance. It may be the case that some assessment criteria in his desired outcomes are more important than others. In other words, the risk assessing agent may regard some assessment criteria as being so crucial that if they are not fulfilled, then this may result in a negative outcome in the business association. On the other hand, it could also be the case that there are some assessment criteria which have a minimal impact on determining the outcome of the interaction. The importance of each assessment criterion in an interaction may vary because of the influence that it has on the outcome of the interaction [1]. Hence, while forming the expectations, we propose that communicating the significance of each assessment criterion is very important, as they play an important part in formalizing the expectations of the interaction [2].

Let us consider the following example of the interaction scenario of agent 'A' forming an interaction with agent 'B' (a logistics company) in the context of forming a contract to transfer goods from Milan to Perth [2]. Let us consider that the expectations of the business contract were formed as follows:

- timely pickup of the goods from agent 'A''s address in Milan on 26 March 2010 (C1)
- meeting the required connections to leave Italy to meet the delivery time in Perth (C2)
- delivery of the goods at agent 'A''s address in Perth on or by 2 April 2010 (C3)

- completion of the business activity in AUD $6000. Out of this, agent ‘A’ has to pay AUD $4000 when the goods are picked up and the remaining AUD $2000 on delivery (C4)

Let us consider that the Easter holiday season in 2010 starts on 3 April 2010 and agent ‘A’ wants his goods to be delivered before this. So, in the above expectations, he may consider assessment criterion C3 to be an important one compared to the other assessment criteria. In the business activity, if the risk assessed agent does not fulfill assessment criterion C3, according to the expectations but fulfills others, then this may lead to agent ‘A’ experiencing a negative outcome and therefore assigning a high risk value to agent ‘B’. Other scenarios may be possible where risk assessing agent ‘A’ may want agent ‘B’ to pick up the goods at the specified time and hence may give more importance to this criterion. In this scenario, if this assessment criterion is not completed, then agent ‘B’ might be assigned a high risk value as the result of a negative outcome in its business association, as shown in Figure 4.5.

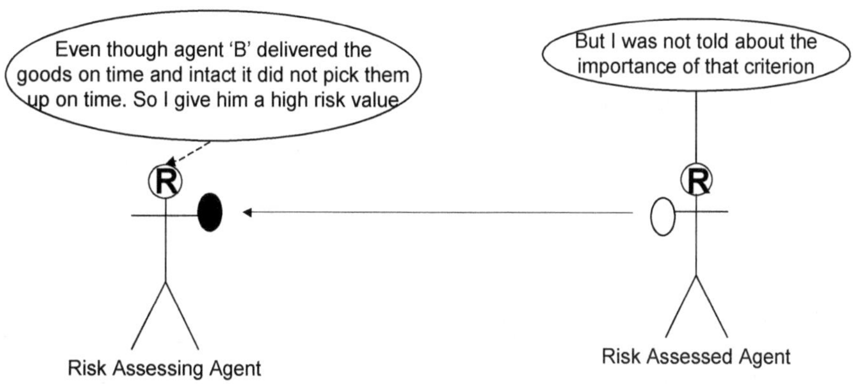

Fig. 4.5 The significance of the assessment criteria

So, depending on the risk assessing agent’s preferences, the significance of each assessment criterion in his expectations on the successful completion of his business activity varies. To have a fair process of the transactional risk analysis, we propose that it is important that the risk assessing agent communicates the significance of the assessment criteria clearly to the risk assessed agent before the process is commenced. This will be important in the next steps of risk assessment, risk communication and risk management, as explained in the next few subsections.

4.3.1 Significance of Assessment Criteria during Risk Assessment Process

Let us consider the business interaction scenario of risk assessing agent ‘A’ who wants to achieve his desired outcomes in a business association. Agent ‘A’ has to make a decision to either form an interaction with risk assessed agent 'B1' or 'B2' to achieve the desired outcomes, of which there are three assessment criteria (termed C1, C2 and C3). Agent 'A' may consider assessment criteria C1 as vital to the successful achievement of his business interaction. By considering agent 'B1's past history, the likelihood of him committing to criteria C2 and C3 according to the expectations of agent 'A' may be high but there may also be a high probability of failure in committing to criterion C1. This may not be the case with agent 'B2', who can commit to assessment criteria C1 with less probability of failure compared to agent 'B1' and criteria C2 and C3 with a greater probability of failure compared to agent 'B1'. In such a case, if the significance of the assessment criteria is not considered, even though the overall level of risk in forming an interaction with agent 'B1' may seem lower compared to agent 'B2', in reality, it may be higher due to the greater level of non-commitment of assessment criterion C1 by agent 'B1', which is considered important by the risk assessing agent. The correct level of risk is determined by weighting the probability of failure of achieving each assessment criterion, according to the significance it will have on the successful completion of the business activity.

4.3.2 Significance of Assessment Criteria during Risk Communication Process

Let us suppose that in the previous case, agent 'A' forms a business association with agent 'B1'. After agent 'A's interaction with agent 'B1', another risk assessing agent 'D' may solicit for recommendations from agent 'A' about risk assessed agent B1's capability in committing to assessment criteria C1, C2 and C3. However, agent 'D' may consider criteria C2 and C3 as significant to the completion of his business interaction. Subsequently, the level of risk communicated by agent 'A' to 'D' in relation to agent 'B1' may be too focused on assessment criterion C1 and not according to what is required by agent 'D'. This might give an incorrect impression of agent 'B1' to 'D' as he considers assessment criteria C2 and C3 vital to the successful completion of its business interaction. This is addressed by communicating the significance of each assessment criterion clearly, along with the recommendations for agent 'D' to make his own assessment during transactional risk analysis.

4.3.3 Significance of Assessment Criteria during the Risk Management Process

The process of risk management is also dependent on the significance of each assessment criterion. In the above case of agent 'A's interaction with agent 'B1', the process of risk management after the risk assessment and risk evaluation process will focus more on assessment criterion C1 than on C2 or C3 (depending on the output of the risk evaluation phase), as it is considered vital for the successful completion of the business activity. Similarly, the focus of risk management may vary in agent 'D's interaction with 'B1' as he considers criteria

C2 and C3 more significant than criterion C1. So, the significance of the assessment criteria also plays an important part in shaping the process of risk management.

In the next section we discuss the process of determining the significance of the assessment criteria in a business activity.

4.4 Formalizing the Significance of the Assessment Criteria

The existing quantitative approaches for risk assessment in the literature use a numerical range of either between 0 and 1[3, 4] or between 0 and 5 [5] to assign a significance level to the assessment criteria of the expectations. In this range, some approaches define different levels of significance; and a risk assessing agent assigns a criterion with a corresponding significance value, according to the level of importance that a criterion has in the successful completion of his business activity. However, there is no systematized process that assists the risk assessing agent in determining the significance value of each assessment criterion. Important factors such as the level of interdependence between the assessment criteria and the relative cumulative importance and impact of each criterion to the goal of the business interaction are ignored, thereby the significance of each assessment criterion by those techniques is assigned in an ad-hoc way. This is similar to those approaches that use a qualitative scale to determine the significance of the assessment criteria [6]. In this book, we propose that the significance of the assessment criteria should be formalized in a systematized way by capturing all the various dependencies or inter-dependences between the assessment criteria. To do this, we propose utilizing Analytic Network Process in our approach which is presented in the next sub-section.

4.4.1 Analytic Hierarchy Process for Determining the Significance of the Assessment Criteria

The Analytic Hierarchy Process (AHP) is a quantitative technique that assists in making structured decisions about the weights to be given to multiple criteria with respect to the goal to be achieved. It achieves this by implementing a pair-wise comparison process among the different assessment criteria and transforming the analysis to numerical representation by which the relative significance of assessment criteria are determined [7]. AHP has been applied in several applications like Balanced Scorecard [8, 9], Decision-Making [10] Geography Information System (GIS) [11], customer-driven product design [12] etc in different domains including education, business, social sciences and natural sciences to determine the importance of different factors with respect to the goal of analysis. AHP requires a hierarchy to be built among the criteria whose significance values have to be determined with respect to the goal. It structures the decision-making problem by calibrating the hierarchies in the problem, for example by goals, criteria, sub-criteria and alternatives [9]. In our problem, to build the hierarchy among criteria the goal to be achieved has to be decomposed into different sub-factors. This is achieved by the risk assessing agent forming the expectations of his interaction with the risk assessed agent which has the assessment criteria shown in Figure 4.6.

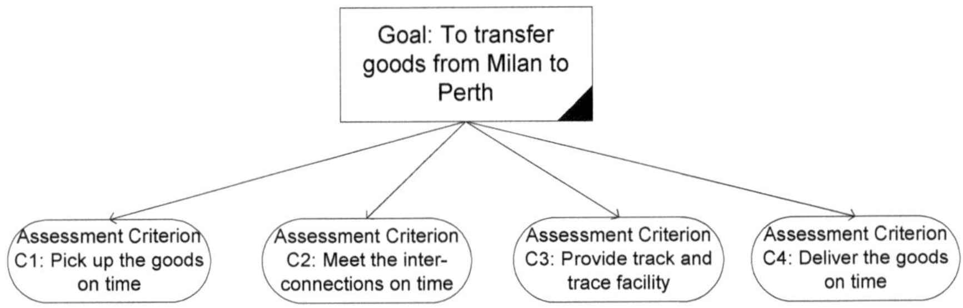

Fig. 4.6 Assessment criteria with respect to the goal of the interaction

Once the hierarchy of the assessment criteria has been formed, a pair-wise comparison between them should be carried out to determine the significance of each with respect to the goal of the business activity. The process of pair-wise comparison among the criteria can be carried out by two approaches, namely conventional and fuzzy analysis. Conventional analysis requires the selection of arbitrary values from a given scale during a pair-wise comparison of the assessment criteria. This approach has a degree of uncertainty associated with it as it is difficult to map qualitative preferences to point estimates [13].This is overcome by using a fuzzy analysis of pair-wise comparisons, which is more appropriate and effective than conventional AHP to capture the uncertainties during pair-wise comparisons. In our approach, we will utilize the fuzzy-based method to determine the weights of the assessment criteria. The series of steps to be followed are:

1. define a range as shown in Figure 4.7 to classify the different levels of importance and significance.
2. define a triangular fuzzy scale as shown in Table 4.1 to determine the different fuzzy sets of importance within the fuzzy scale.
3. determine the triangular fuzzy reciprocal value for the different linguistic fuzzy sets defined in step 2.
4. construct an nXn matrix from pair-wise comparisons of the different assessment criteria with respect to the final goal to be achieved in the business activity. ‘n’ represents the number of assessment criteria in the business activity.
5. utilize the fuzzy synthetic extent approach to determine the degree of possibility of a given assessment criterion being more important than others.
6. the analysis of step 5 will represent the degree of importance of the different assessment criteria with respect to the goal of the interaction.

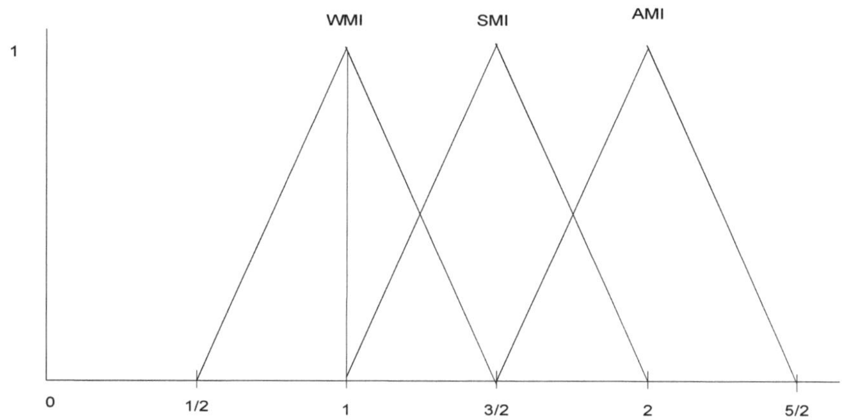

Fig. 4.7 Linguistic scale for measuring the relative importance of an assessment criterion

Table 4.1 Linguistic scale for importance of the assessment criterion

Linguistic Scale for Importance	Triangular Fuzzy Scale	Reciprocal Fuzzy Value
Just Equal	(1,1,1)	(1, 1, 1)
Weakly more Important	(1/2, 1, 3/2)	(2/3, 1/2, 2)
Slightly more Important	(1, 3/2, 2)	(1/2, 2/3, 1)
Absolutely more Important	(3/2, 2, 5/2)	(2/5, 1/2, 2/3)

Extending our example from Section 4.3, let us consider agent 'A' wants the goods to be delivered at the destination address by 2nd April 2010 and considers this assessment criterion as the most significant and important one for the successful achievement of his business activity. Determining the pair-wise comparison of each assessment criterion of Figure 4.6 with respect to the other assessment criteria, we obtain the 4 X 4 comparison matrix, as shown in Table 4.2. Using the fuzzy synthetic approach, the weights of each assessment criteria determined are shown in the last column of Table 4.2.

Table 4.2 Pair-wise comparison matrix and the weights of the assessment criteria

Assessment Criteria (C)	C1	C2	C3	C4	Weight
C1	(1,1,1)	(3/2,2,5/2)	(3/2,2,5/2)	(2/3,1/2,2)	0.377
C2	(2/5,1/2, /3)	(1,1,1)	(1/2,2/3,1)	(2/5,1/2,2/3)	0.041
C3	(2/5,1/2,2/3)	(1,3/,2)	(1,1,1)	(2/5,1/2,2/3)	0.178
C4	(1/2,1,3/2)	(3/2,2,5/2)	(3/2,2,5/2)	(1,1,1)	0.404

From the analysis, it can be seen that assessment criterion C4 is the most important of the expectations followed by C1, C3 and C2. The weights of these assessment criteria are determined by the risk assessing agent in comparison with the other criteria and hence the analysis gives an informed representation of their importance to the successful completion of the business interaction. Existing approaches may have given the highest level of significance to assessment criteria C4, but they do not make any informed representation on the significance value that has to be assigned to the other assessment criteria C1- C3. This is addressed by using AHP to ascertain the significance of each assessment criterion.

4.4.2 Limitations of Analytic Hierarchy Process in Determining the Significance of the Assessment Criteria

One limitation of using the Analytic Hierarchy Process (AHP) to find the weights of the assessment criteria is that it only works well when there is no inter-dependence between the assessment criteria [14]. In other words, the AHP process works well when all of the assessment criteria are independent of the others and do not have an effect on the successful achievement of the other assessment criteria. For example, consider the following objective of forming a business association as shown in Figure 4.8. In such an activity, each assessment criterion is important to the successful achievement of the business goal but none of them have an effect on the successful achievement of the other assessment criteria. In other words, there is no implied relationship between them.

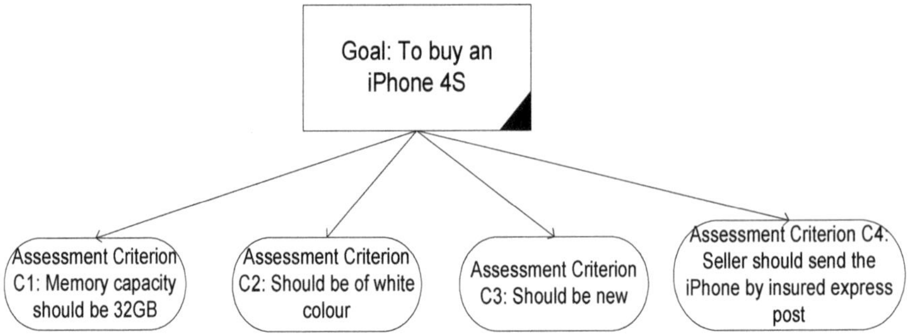

Fig. 4.8 Expectations and assessment criteria without dependency between them

In contrast, there may be certain business activities where the assessment criteria are inter-dependent. For example, considering the logistics interaction scenario mentioned in Figure 4.6, in order for the goods to be delivered on time a sequence and series of steps need to be completed beforehand, such as the goods being picked up on time, and the different transport connections be met before the deadline which ensures that the goods reach the destination before the holiday closedown period. Also, let us consider that for the logistic company to activate the track and trace facility the goods should reach the logistic company's warehouse from the remote location when they are picked up, as shown in Figure 4.9.

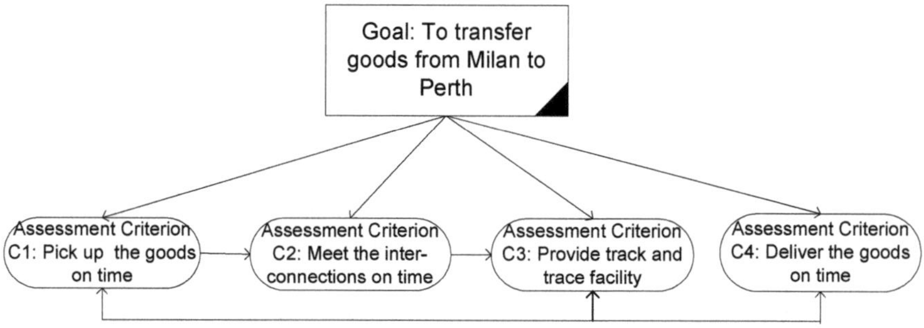

Fig. 4.9 Assessment Criteria with respect to the goal and the inter-dependencies between them

On a scale of either [0-1] or [0-5], the risk assessing agent may assign the highest importance value to the assessment criterion 'deliver the goods on time' (C4), as he considers this to be the most important assessment criterion. This, however, does not take into consideration the level of dependency that assessment criteria C1and/or C2 have on assessment criterion C4, in relation to Agent 'B' being able to meet that criterion while assigning the significance value for them. For example, if the first assessment criterion of picking up the goods on time (C1) is not met, then it is quite difficult for agent 'B' to meet the connections in time (C2) which will, in turn, have an effect on delivering the goods on time (C4). The risk assessing agent assigning a very high significance value to assessment criterion C4 and a low significance value to criteria C1 and C2 does not capture the level of dependency that C4 has on those criteria. Thus, a technique is needed which, in addition to considering the importance of each assessment criteria with respect to the goal, also considers the level of dependencies between the different assessment criterion and ascertains their significance values accordingly. To address this in our approach, we utilize the Analytic Network Process (ANP) to ascertain the different weights of the expectations, according to the inner dependencies between them. The approach is presented in the next sub-section.

4.4.3 Analytic Network Process to Determine the Significance of the Assessment Criteria

The Analytic Network Process (ANP) determines the impact of a given set of factors on the goal to be achieved by considering the levels of interdependencies between them. This is in contrast to its predecessor, Analytic Hierarchy Process (AHP), which only considers the weight of the factors on the goal. Many decision-making problems cannot be structured hierarchically because they involve the interaction and dependence of higher-level elements on lower-level elements [15]. When network system relationships exist between the criteria, the ANP can be used to make a better decision. ANP models the relationships between the criteria in two parts: the first models the control hierarchy among the criteria according to the goal, whereas the second models the different sub-networks of influences among the criteria, for each criterion [16]. The inter-relationships between the decision levels and attributes are modeled in ANP by determining the composite weights through the development of a supermatrix [7]. The supermatrix is a partitioned matrix, where each matrix segment represents a relationship between two components or clusters in a system

As with AHP, the level of inter-dependence between the criteria can be modeled by conventional or fuzzy techniques. In our approach, we use the fuzzy ANP process, which can tolerate vagueness and ambiguity in the information, thereby assisting in the decision-making process. The steps for the Analytic Network Process are an extension of the AHP steps mentioned in section 4.4.1. At first, the pair-wise comparison of each assessment criterion with respect to the others is determined and their importance in relation to the goal is ascertained, as shown in Table 4.2. Then, the following extension steps are performed as follows [14]:

1. determine the level of dependencies among the different assessment criteria and compute a pair-wise comparison matrix for each dependant criterion, according to the linguistic scale defined in step 2 of the AHP
2. repeat steps 2- 6 mentioned in the previous sub-section to obtain the relative importance weights for each inter-dependence matrix
3. multiply the weights obtained in step 6 of the AHP with the ones obtained in step 2 of the ANP to ascertain the final weights of each assessment criterion with respect to the goal and the inter-dependencies between them.

We illustrate the working of these steps with an example using the interaction scenario mentioned in Figure 4.9 and the inter-dependencies between the assessment criteria, as shown in Figure 4.10. It can be seen from the figure that assessment criteria C4 is dependent on C1 and C2, C3 is dependent on C2 and C1, C2 is dependent on C1 and there is an inter-dependency between C2 and C4.

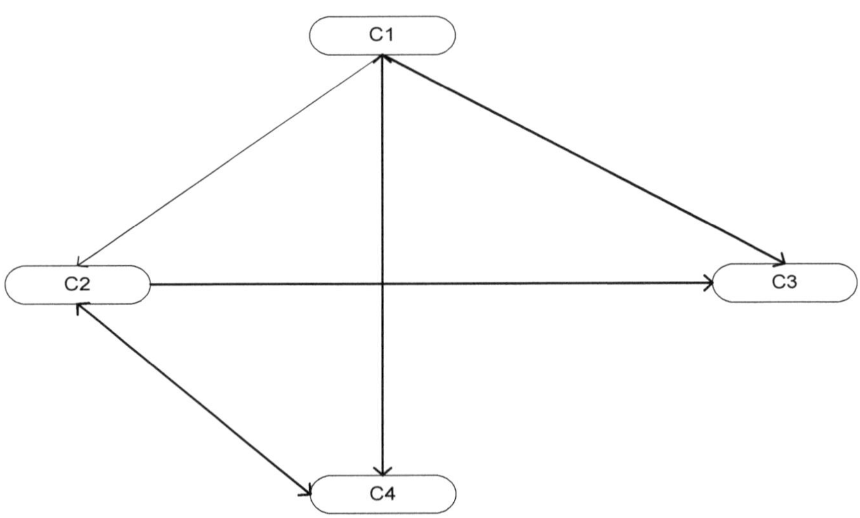

Fig. 4.10 Level of dependency between the assessment criteria of the logistics interaction

Once the importance of the assessment criteria with respect to the goal (shown in Table 4.2) is determined, then the inner dependence matrix of the assessment criteria should be ascertained using pair-wise comparisons. The inner dependence matrices are shown in Tables 4.3 –4.5.

Table 4.3 Showing the inner dependence matrix for assessment criterion C4 on other criteria

Inner dependence matrix with respect to assessment criterion C4			
C4	C1	C2	Weight
C1	(1,1,1)	(1/2,2/3,1)	0.314
C2	(1,3/2,2)	(1,1,1)	0.685

Table 4.4 Showing the inner dependence matrix for assessment criterion C2 on other criteria

Inner dependence matrix with respect to assessment criterion C2			
C2	C1	C4	Weight
C1	(1,1,1)	(3/2,2,5/2)	0.819
C4	(2/5,1/2,2/3)	(1,1,1)	0.180

Table 4.5 Showing the inner dependence matrix for assessment criterion C3 on other criteria

Inner dependence matrix with respect to assessment criterion C3			
C3	C1	C2	Weights
C1	(1,1,1)	(1,3/2,2)	0.685
C2	(1/2,2/3,1)	(1,1,1)	0.314

Multiplying these weights with the individual weights determined in Table 4.2, we get the final weights of the assessment criteria as follows:

Table 4.6 Significance values of the assessment criteria after considering the interdependencies

Assessment Criteria	Previous Significance Value	Significance value after considering inter-dependencies
C1	0.377	0.328
C2	0.041	0.1867
C3	0.178	0.089
C4	0.404	0.3963

As can be seen from Table 4.6, the significance values of the assessment criteria change considerably after considering the interdependencies between them. The importance of the assessment criteria in descending order is now C4, C1, C2 and C3 as compared to C4, C1, C3 and C2, previously. Thus, it can be concluded that assessment criterion C2 is an important step for agent 'B' to commit to as compared to criterion C3 if agent 'A' has to achieve his 'goods delivered on time at the destination address' criterion thereby achieving a neutral outcome. ANP helps to consider such inter-dependencies among the assessment criteria and correspondingly determines their significance values with respect to the goal of the interaction.

In the next section, we discuss the division of time according to which the dynamic nature of transactional risk can be considered during the risk identification and risk assessment steps.

4.5 Division of Time to Capture the Dynamicity of Transactional Risk

There are different time scenarios that could be possible in the risk assessing agent's interaction with the risk assessed agent. For example, in the interaction scenario mentioned in Figure 4.11, it may be possible that risk assessing agent 'A' expects his interaction with agent 'B' to last a week. In some scenarios, it is possible that the business association between the agents will last for a longer period of time (say one month or six months), as shown in Figure 4.12.

Fig. 4.11 Duration of agent 'A''s interaction with agent 'B'

Fig. 4.12 Duration of agent 'A''s interaction with agent 'B'

In each case, the duration of the interaction time varies, and with that, the length of time in which perceived transactional risk has to be identified and analyzed also varies. As discussed in Section 3.3.3, such variations in time scenarios have to be considered while ascertaining perceived risk in accordance with its dynamic nature. To capture the dynamic nature of transactional risk while undertaking risk assessment, in the previous chapter, we introduced concepts such as time space, time slot and time spot to capture the time period while conducting transactional risk analysis. Based on these terms, we adopt the methodology proposed in Chang et al. [1] which utilizes the following approach:

(a) The risk assessing agent should first determine the total duration of time over which he wishes to analyze and ascertain the perceived transactional risk in interacting with the risk assessed agent. This total duration of time is termed the *time space* (as discussed in section 3.2.12). In other words, as shown in Figure 4.13, the time space of the interaction is a finite duration of time over which the risk assessing agent wishes to ascertain and analyze the perceived transactional risk by determining its subcategories, and subsequently make a risk-based decision in forming an interaction with the risk assessed agent.
(b) The risk assessing agent should then divide the time space into a finite number of non-overlapping, mutually exclusive and equally spaced sectors of time. Each equally spaced sector of time is termed the *time slot* (as discussed in section 3.2.13). In other words, as shown in Figure 4.13, the whole time space can be regarded as a series of a finite number of non-overlapping, mutually exclusive time slots.
(c) The time at which the risk assessing agent interacts or plans to interact with a risk assessed agent is termed the *time spot* (as discussed in section 3.2.14).

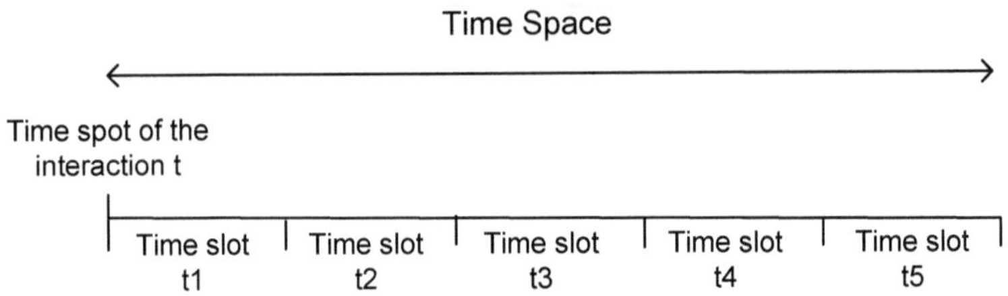

Fig. 4.13 Duration of time for transactional risk assessment

To capture the dynamic nature of transactional risk, in this book we consider that the risk assessing agent should determine each subcategory of transactional risk, namely performance risk and financial risk in each time slot of his business activity. As mentioned in the earlier chapters, performance risk represents the inability of the risk assessed agent to commit to the assessment criteria as decided in the expectations in each time slot of its time space. Determining the performance risk in each time slot will capture variations in the risk assessed agent's ability to act according to the expectations at different periods of time of the interaction. However, it is possible that the risk assessing agent's interaction with the risk assessed agent may be at a point of time in the future, as shown in Figure 4.14.

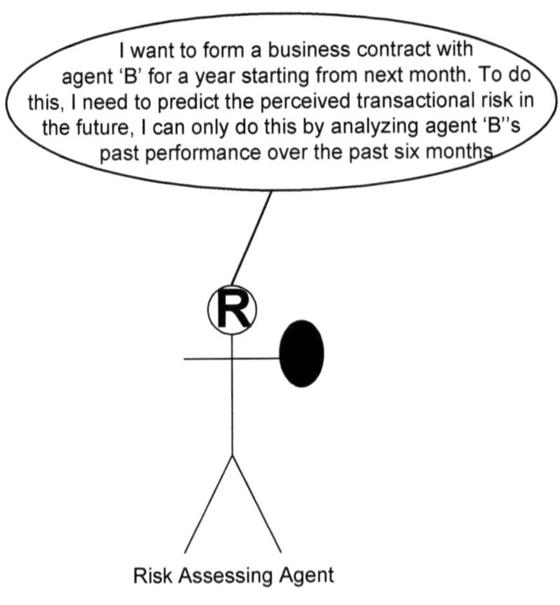

Fig. 4.14 Duration of time for the analysis of performance risk of agent 'B'

In such cases, the risk assessing agent, in ascertaining the performance risk of the risk assessed agent, has to utilize prediction methods and model the inability of the risk assessed agent to commit to the expectations, over points of time in the future. This can only be achieved if the risk assessing agent has a sufficient previous impression or knowledge of the risk assessed agent in relation to the required assessment criteria, as shown in Figure 4.14. By using such past impressions or knowledge, the risk assessing agent can utilize the prediction techniques to model the risk assessed agent's performance risk over a period of time in the future. However, to consider scenarios such as these, it is important that the risk assessing agent models the time period of the business interaction (shown in Figure 4.13), so that it takes into account:

- the previous duration of time before the time spot of its interaction, in which he considers the previous impression of the risk assessed agent, according to the expectation of his future interaction;
- the present or future duration of time after the time spot of his interaction, in which he actually interacts with the risk assessed agent, and subsequently, wants to determine the perceived transactional risk in this time period.

After choosing the duration of the time space of the interaction, we propose that the risk assessing agent should divide it into two broad phases, as shown in Figure 4.15. These two phases are:

1. pre-interaction start time phase
2. post-interaction start time phase

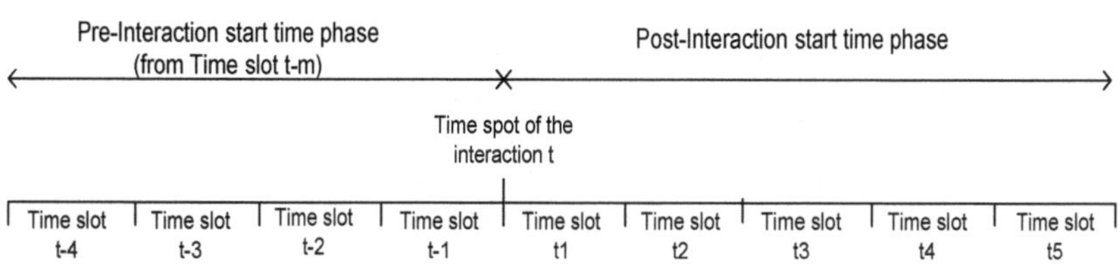

Fig. 4.15 Division of the time space into the two time phases

4.5.1 Pre-interaction Start Time Phase

The pre-interaction start time phase of the time space is termed as that period of time before the risk assessing agent starts his interaction with the risk assessed agent. The pre-interaction start time phase spans the time from the beginning of the time space to the time spot of the interaction, as shown in Figure 4.15.

4.5.2 Post-interaction Start Time Phase

The post-interaction start time phase of the time space is termed as that period of time after the risk assessing agent initiates and interacts with the risk assessed agent. Such a time phase spans from the time spot of the interaction to the end of the time space as shown in Figure 4.15.

Once the time space of the business activity has been formed then the risk assessing agent can initiate the process of risk assessment. We discuss about the process further in the next chapter.

4.6 Recapitulation

In this chapter, we presented the factors that need to be considered during the process of risk identification in business activities. We started by explaining the process of formalizing the assessment criteria and expectations in the two different types of business activities. We then explained the importance of formalizing the significance values of the assessment criteria of the expectations by considering the various inter-dependencies between them. We then proposed an approach by which the time space of the business activity has to be formed to capture the dynamic nature of transactional risk. A brief overview of the different steps to be carried out in each of these factors was presented. Once this process is over, the risk assessing agent can identify the different events which can lead to risk and then ascertain the subcategories of performance risk and financial risk in each time slot to determine the perceived transactional risk in the business activity. We will discuss further the process of risk assessment in Chapters 5 and 6.

References

1. Chang, E., Dillon, T., Hussain, F.K.: Trust and Reputation for Service-Oriented Environments. John Wiley & Sons, Ltd., West Sussex (2006)
2. Hussain, O., Sangka, K.B., Hussain, F.K.: Determining the Significance of Assessment Criteria for Risk Analysis in Business Associations. In: Lu, J., Jain, L.C., Zhang, G. (eds.) Handbook on Decision Making, vol. 33, pp. 403–416. Springer, Heidelberg (2012)
3. Wang, Y., Wong, D.S., Lin, K.-J., Varadharajan, V.: Evaluating transaction trust and risk levels in peer-to-peer e-commerce environments. Information Systems and E-Business Management 6, 25–48 (2008)
4. Hussain, O., Dillon, T.: Cost-Benefit Analysis to Hedge with Third-Party Producers in Demand-Driven Production. In: Xhafa, F., Barolli, L., Papajorgji, P.J. (eds.) Complex Intelligent Systems and Their Applications, vol. 41, pp. 69–81. Springer, New York (2010)
5. Hussain, O.K., Chang, E., Hussain, F.K., Dillon, T.S.: A methodology to quantify failure for risk-based decision support system in digital business ecosystems. Data & Knowledge Engineering 63, 597–621 (2007)
6. Hussain, O.K., Dillon, T., Hussain, F., Chang, E.: Probabilistic Assessment of Loss in Revenue Generation in Demand-Driven Production. Journal of Intelligent Manufacturing (2011), doi:10.1007/s10845-011-0518-4
7. Saaty, T.L.: The Analytic Network Process (2008)
8. Searcy, D.L.: Aligning the Balanced Scorecard and a Firm's Strategy Using the Analytic Hierarchy Process. Management Accounting Quarterly 5, 1–10 (2004)

9. Jovanovic, J., Krivokapic, Z.: AHP In Implementation Of Balanced Scorecard. International Journal for Quality Research 2, 59–67 (2008)
10. Wu, H.-H., Shieh, J.I., Li, Y., Chen, H.-K.: A Combination of AHP and DEMATEL in Evaluating the Criteria of Employment Service Outreach Program Personnel. Information Technology Journal 9, 569–575 (2010)
11. Duc, T.T.: Using GIS and AHP Technique for Land-use Suitability Analysis. In: International Symposium on Geoinformatics for Spatial Infrastructure Development in Earth and Allied Sciences, pp. 1–6 (2006)
12. Lina, M.-C., Wanga, C.-C., Chena, M.-S., Chang, C.A.: Using AHP and TOPSIS approaches in customer-driven product design process. Computers in Industry 59, 17–31 (2008)
13. Yu, C.-S.: A GP-AHP method for solving group decision-making fuzzy AHP problems. Computers & Operations Research 29, 1969–2001 (2002)
14. Yuksel, I., Dagdeviren, M.: Using the fuzzy analytic network process (ANP) for Balanced Scorecard (BSC): A case study for a manufacturing firm. Expert Systems with Applications 37, 1270–1278 (2010)
15. Saaty, T.L., Takizawa, M.: Dependence and independence: From linear hierarchies to nonlinear networks. European Journal of Operational Research 26, 229–237 (1986)
16. Saaty, T.L.: Multi-decisions decision-making: In addition to wheeling and dealing, our national political bodies need a formal approach for prioritization. Mathematical and Computer Modelling 46, 1001–1016 (2007)

Chapter 5
Risk Assessment Phase: Performance Risk Assessment in Business Activities

5.1 Introduction

In the previous chapter, we explained the steps for risk identification by which the risk assessing agent formulates the scope and context of his business activity. The next step in the process of risk analysis is to assess the level of perceived transactional risk. In Chapter 2, we noted that transactional risk in a business association is a combination of two subcategories, namely performance risk and financial risk. Performance risk in a business activity ascertains the chances of the activity failing due to the dependent expectations of the risk assessing agent not being met as decided initially. However, it does not consider the likely impact which is associated with the transactional risk. In this chapter, we consider all factors that characterize the perceived transactional risk which was discussed in section 3.3 and propose an approach and the mathematical model to determine the performance risk in a business activity.

5.2 Importance of a Unified Measure for Ascertaining Performance Risk

In Section 3.5 we discussed the two ways of initiating the process of risk assessment, namely basing it on the risk assessing agent's own experiences or soliciting recommendations. We also discussed in Section 3.6 that a risk assessing agent may either have a single risk relationship or multiple risk relationships with a risk assessed agent that have been established through his previous associations with that agent in either the context of same or different interactions. The risk set which is formed as an outcome of each association details the measure of perceived transactional risk and its subcategories. During performance risk assessment, the risk set represents to the risk assessing agent (say agent 'A') the likely level of failure to achieve the dependent outcomes due to the non-performance of the risk assessed agent, as decided in the contracted expectations. This information is utilized further when the risk assessing agent in question has to either:

O.K. Hussain et al.: Risk Assessment & Management in the Networked Economy, SCI 412, pp. 107–150.
springerlink.com

(a) utilize the information to ascertain the performance risk in engaging in a future business association with the risk assessed agent, or
(b) provide recommendations for the risk assessed agent to any other agent when these are solicited.

In both cases, it is important that a clear and unambiguous representation be used to avoid confusion and misinterpretation of information.

For example, let us consider two scenarios of a risk assessing agent 'A' wanting to enter into a business association with agent 'B' to buy an iPhone4S. The first scenario is that agent 'A' has had a previous association with agent 'B' as shown in Figure 5.1. The second is when agent 'A' solicits recommendations from other agents and receives a reply as shown in Figure 5.2.

Fig. 5.1 The risk assessing agent 'A' having his own interaction history with agent 'B'

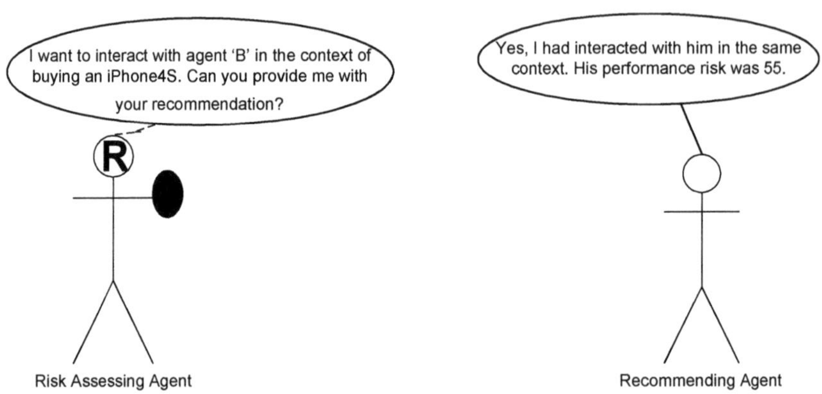

Fig. 5.2 The risk assessing agent 'A' solicits recommendations about agent 'B' from other agent

The question that arises in the first scenario is 'what does the risk assessing agent mean by *reliable*'? and in the second scenario is 'what does the recommending agent mean by the value *55*'? This problem is further exacerbated by the fact that different agents can utilize different terms to express their opinions (for example 'good', 'OK' etc). At a later time, the same agent or different agents may have difficulty interpreting the meaning of these terms. For example, in the first scenario, agent 'A' termed agent 'B' as reliable in the previous business association. But after five years, agent 'A' may be unsure what he meant by the term reliable at that point in time. In order to avoid such situations, there is a need for a standardized representation that will be used as a basis to represent performance risk. Such a basis will have a common understanding of all the agents and will be utilized to assign and to have a unified representation of the performance risk of different agents. This will eliminate any ambiguities in understanding by different agents. To have such a standardized representation, a scale and quantity (numerical or linguistic) measure should be used. We discuss this further in the next section.

5.3 Failure Scale

To minimize the confusion which arises by using arbitrary and ambiguous measures to represent the performance risk in a business activity, in this book we use a non-overlapping numeric scale in the range of 0 to 100. This scale is known as the *Failure Scale* and it represents the different levels of non-commitment in a business activity. Having such a scale gives a domain, a platform on which the risk assessing agent or the recommending agent can represent in clear and concise terms the level of non-commitment of the risk assessed agent to the dependent events of the expectations. Furthermore, each value on this scale represents varying levels of non-commitment that can be utilized to represent both the severity, magnitude of failure or non-commitment by the risk assessed agent to the expectations in the business activity. Approaches have been proposed in the literature that determine and classify either the feedback, or trust or risk values in different continuous ranges between -1 to 1, 0 to 1 or -1 to 0. However, one of the drawbacks of these approaches is that such a range is not expressive in classifying and clearly expressing the severities for the risk assessed agent. Using a scale of 0-100 represents varying levels that can be used to more accurately capture and represent clearly the different levels of failure.

However, even though such a scale is used to represent the risk assessed agent's level of non-commitment, it is still necessary to make sense of the values in that scale. For example, let us consider the interaction scenario discussed in Figure 5.2 where the recommending agent summarizes the performance risk of agent 'B' as 55. The question that arises here is: what does the value '55' in the range of 0-100 signify? Does it signify that the performance risk in the business activity leans towards the higher level, considering that 0 is the least level of performance risk and 100 the highest level, or vice versa? If this is the case, then how extreme or severe is it? Such a scenario is further complicated when the risk assessing agent has two or more recommending agents providing their recommendations to him in different numeric values as shown in Figure 5.3.

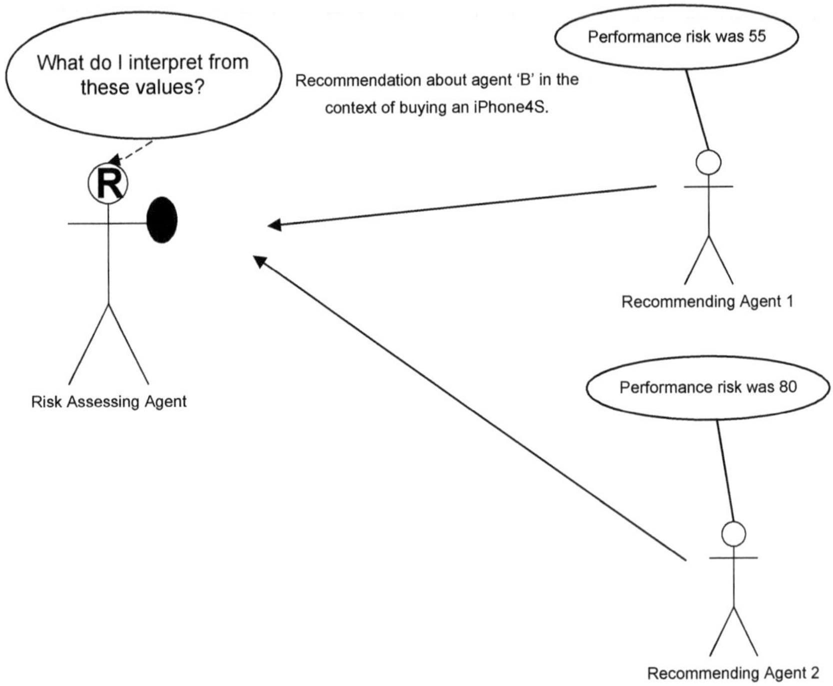

Fig. 5.3 Agent 'A' receiving recommendations from two recommending agents

So to make sense of the different measures of severity and magnitude of failure, it is important to add semantics to the different values in that range. Adding semantics to a numerical value classifies it according to its range and enables a better understanding of its representation. All the approaches which consider the trust or risk values of an agent in different continuous ranges like -1 to 1, 0 to 1 or -1 to 0 do not have associated semantics thereby making it difficult to interpret the meaning of a variable within that range. This means that a certain level of ambiguity might be present when the information is propagated between different agents or is provided by different agents.

However, in defining semantics, it is important to consider the total numeric range and identify various semantic tags that define the level of severity of the numeric value or range. For example, there are approaches proposed in the literature which use the two extreme adjectives of "High" or "Low" as the semantic tags to represent a variable within a range. But as shown in Figure 5.4, these tags do not help in any way to alleviate the ambiguity in the risk assessing agent's mind in relation to the performance risk.

So, when defining semantics, it is important to consider the total numeric range, divide them into different levels according to the level of severity to which they correspond, and associate them with a semantic tag that defines the associated magnitude of severity. This also helps to add semantics to a group of values in a range and not to each value in the numeric range, which is a tedious process.

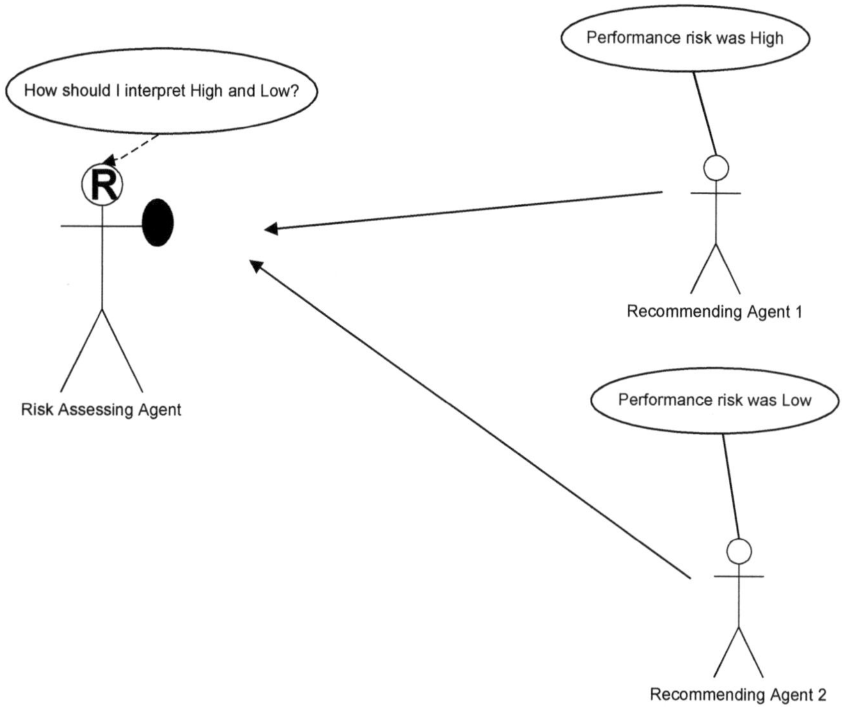

Fig. 5.4 Agent 'A' receiving recommendations from two recommending agents in semantic terms

In the next section, we divide the Failure Scale into different levels depending upon the severity, and define the semantic tags for each level.

5.4 Adding Semantics to the Failure Scale

To add semantics to the Failure Scale, we divide it into seven different levels of failure varying according to their severity as shown in Figure 5.5 [1]. These levels of failure effectively show the different levels of non-commitment of the risk assessed agent to the dependent events. Each level of failure on the Failure Scale is represented both by a numeric value and semantic tag which shows the corresponding severity of failure associated with it. The numeric value is termed as the 'FailureLevel' (FL). In other words, FailureLevel quantifies the different levels of failure according to the range of its severity on the Failure Scale.

The level of failure in interacting with a risk assessed agent at a given time 't' which can be either in the past, current or future time period is represented on the Failure Scale with the metric FailureLevel (FL). As shown in Figure 5.5, the domain of the Failure Scale ranges from [-1, 5]. We define the domain of the Failure Scale as the possible set of values from which a FailureLevel is assigned to

Semantics of Failure Level	Level of non-commitment	FailureLevel
Unknown	-	-1
Total Failure	91-100 %	0
Extremely High	71-90 %	1
Largely High	51-70 %	2
High	26-50 %	3
Significantly Low	11-25 %	4
Extremely Low	0-10 %	5

Fig. 5.5 The Failure Scale

the risk assessed agent, according to the severity of failure present in interacting with it, at a given period of time. We choose to have a single integer representing the FailureLevel, for the range of severity of failure covered by it. The reasons for choosing this domain to represent the FailureLevel of a risk assessed agent are:

- As mentioned earlier, some work in the literature has been done which represents the possible risk in an interaction just as a result of performance risk only within a domain of [0, 1]. However, this domain for representation is not very expressive. In our method, we use a domain which alleviates the abovementioned disadvantages and is therefore more expressive and can represent different levels of failure according to their severity.
- It provides one level to represent the state of ignorance in the magnitude of failure in the business activity or interaction (Level -1).
- Two levels to represent high magnitude of failure in an interaction (Level 0 and 1). Of these two levels, one represents the greater level of high magnitude of failure, and the other represents the lesser level of high magnitude or degree of failure in an interaction.
- Two levels to represent the medium magnitude of failure in an interaction (Level 2 and 3). Of levels, one represents the higher level of medium magnitude of failure and the other level represents the lower level of medium magnitude of failure in the interaction.
- Two levels to represent low magnitude of failure in an interaction (Level 4 and 5). One level represents the higher level of low magnitude of failure and the other level represents the lower level of low magnitude of failure in the interaction.

Hence, the domain that we propose for the Failure Scale ranges from [-1, 5], with -1 representing the level of failure as 'Unknown' and the levels from 0 to 5 representing the decreasing range of severity of failure. In order to express semantically each level of failure on the Failure Scale, we define the semantics or meanings associated with each FailureLevel as follows.

5.4.1 Unknown

The first level of the Failure Scale is termed the *Unknown* level of failure or *Unknown* level of non-commitment and its corresponding FailureLevel is -1. This level suggests that the level of failure in interacting with the risk assessed agent is unknown.

Semantics: This level can be assigned only by the recommending agent to the risk assessed agent if he does not have any past interaction history with him according to the context and assessment criteria against which the recommendation is made. Hence, we propose that the recommending agent, instead of recommending any random FailureLevel in the range of (0, 5) on the Failure Scale, recommends the level -1 to the risk assessing agent soliciting recommendations.

5.4.2 Total Failure

The second level of the Failure Scale is defined as *Total Failure* and its corresponding FailureLevel value is 0. A FailureLevel of 0 suggests that the probability of failure in interacting with the risk assessed agent or level of non-commitment is between 91-100 %.

Semantics: This level on the Failure Scale suggests that at a given point in time and with respect to the given assessment criteria, the risk assessed agent is completely unreliable in terms of completing the outcomes desired by the risk assessing agent. In other words, he does not complete the interaction according to the expectations at all and acts fraudulently in the business activity, thus resulting in the risk assessing agent's total failure to achieve his desired outcomes. The FailureLevel of 0 expresses the highest level of failure possible in the business activity.

5.4.3 Extremely High

Extremely High is the third level on the Failure Scale with a corresponding FailureLevel value of 1. This level denotes that there is 71-90 % probability of failure in interacting with the risk assessed agent.

Semantics: This level on the Failure Scale depicts that at a given point in time and with respect to the given assessment criteria, the risk assessed agent is extremely unreliable in committing to the desired outcomes of the risk assessing agent. In other words, he will deviate from the expectations most of the time, thereby resulting in an extremely high level of failure in the interaction accordingly.

5.4.4 Largely High

The fourth level of the Failure Scale is termed *Largely High* level of failure. The corresponding FailureLevel value of this level is 2. This level indicates that there is a 51-70 % probability of failure in interacting with the risk assessed agent.

Semantics: A FailureLevel of 2 on the Failure Scale indicates that there is a significant high level of failure in the business activity, as the risk assessed agent at that given point of time will not commit to a significantly large extent to the expectations of the interaction.

5.4.5 High

The fifth level on the Failure Scale is termed *High* level of failure, and it is represented by a FailureLevel value of 3. This level outlines that there is 26-50 % probability of failure in the interaction.

Semantics: A FailureLevel value of 3 on the Failure Scale assigned to a risk assessed agent suggests that at that particular point in time, the risk assessed agent is unable to complete the business activity to a large extent according to the expectations, thereby resulting in a high level of failure in the interaction.

5.4.6 Significantly Low

The sixth level on the Failure Scale is defined as *Significantly Low* level of failure with a corresponding FailureLevel value of 4. This level indicates that there is 11-25 % probability of failure in the interaction.

Semantics: This level on the Failure Scale suggest that at a given point of time the risk assessed agent can complete most but not all of the assessment criteria according to the expectations of the business activity. A FailureLevel of 4 on the Failure Scale indicates that the risk assessed agent assigned with this value can be relied on to a greater extent, at that time, to commit to the desired outcomes of the business activity, thus resulting in a significantly low failure level in the interaction.

5.4.7 Extremely Low

Extremely Low is the seventh and the last level of the Failure Scale represented by the FailureLevel value of 5. This level shows that there is 0-10 % probability of failure in the interaction.

Semantics: This level on the Failure Scale implies that at a given point in time, the risk assessed agent can be fully relied upon to complete the business activity according to the expected behavior, thereby minimizing the probability of failure in an interaction. The probability of failure in interacting with the risk assessed agent, if any, will be minimal. A FailureLevel of 5 expresses the lowest level of failure possible in a business activity.

By using the Failure Scale and the semantics defined, the performance risk of the risk assessed agent can be represented both by the risk assessing agent or the recommending agent in clear and unambiguous terms in different time phases of the time space. In the next section, we discuss the different time scenarios in which perceived transactional risk needs to be assessed in a business interaction.

5.5 Classification of the FailureLevel Value according to Time

As mentioned in the earlier section, the performance risk in forming a business association with a risk assessed agent at a given time is represented on the Failure Scale with the metric FailureLevel (FL). However, in Section 4.5 we described the process by which the risk assessing agent forms the time space of the business activity and divides it into two time phases to consider the dynamic nature of transactional risk during its assessment. The process of ascertaining the FailureLevel of the risk assessed agent in each time slot varies according to the time phase in which it occurs. We discuss that further in the next section. However, firstly we classify and differentiate the way how FailureLevel is represented according to the time phase of the business activity in which it is being determined.

5.5.1 Previous FailureLevel (PFL)

As shown in Figure 5.6, if the FailureLevel for a risk assessed agent is determined for the time slots of the pre-interaction start time phase, then it is represented as 'PFL' which stands for 'Previous FailureLevel'. Further explanation of how to determine the PFL is given in Section 5.9.

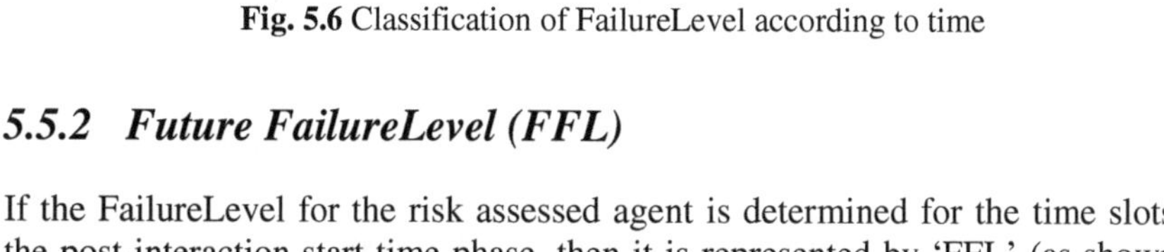

Fig. 5.6 Classification of FailureLevel according to time

5.5.2 Future FailureLevel (FFL)

If the FailureLevel for the risk assessed agent is determined for the time slots of the post-interaction start time phase, then it is represented by 'FFL' (as shown in Figure 5.6) which stands for 'Future FailureLevel'. The process of determining the FFL is discussed in Section 5.12.

5.5.3 Actual FailureLevel (AFL)

The FailureLevel which the risk assessing agent determines for the risk assessed agent after interacting with him is represented by 'AFL' which stands for 'Actual FailureLevel'. This is the FailureLevel determined in the risk recording phase.

Further explanation about determining the AFL of the interaction is presented in Chapter 10.

5.6 Different Time Scenarios of Perceived Transactional Risk Analysis in a Business Activity

In this section, we identify the different time scenarios in which the risk assessing agent's business activity with the risk assessed agent falls and the steps that need to be taken in each of them to ascertain the FailureLevel in forming an interaction with it [2].

5.6.1 Ascertaining the FailureLevel before the Time Spot of the Business Activity

Scenario 5.1: The risk assessing agent has to determine the FailureLevel of the risk assessed agent in a time slot before the time spot of his interaction, i.e. in the pre-interaction start time phase.

The risk assessing agent can determine the FailureLevel (PFL) of the risk assessed agent according to the expectations of his future interaction with him, in a pre-interaction start time slot (t-m) as shown in Figure 5.7 by considering either:

- his previous interaction history with him (if any) in the expectations of his future interaction, (direct past interaction-based probability of failure); or
- in the case of ignorance, then soliciting for recommendations from other agents according to the expectations of his future interaction and assimilating them (reputation-based probability of failure).

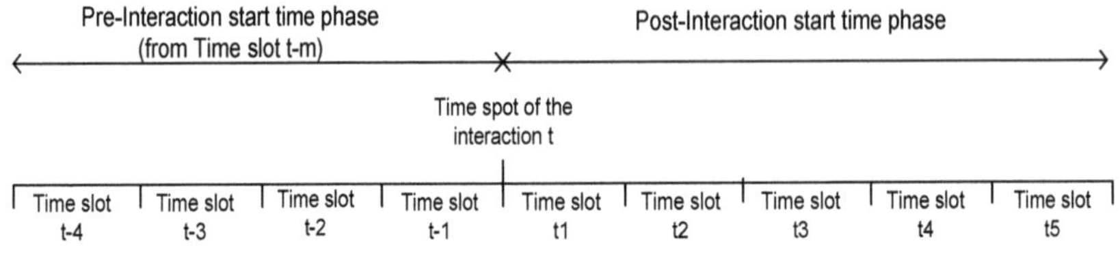

Fig. 5.7 Determining the PFL of a risk assessed agent

A detailed explanation of how to determine the FailureLevel of the risk assessed agent in a time slot by using either direct past-interaction history or by soliciting recommendation from other agents is given in Section 5.9.

5.6.2 *Ascertaining the FailureLevel after the Time Spot of the Business Activity*

Scenario 5.2: The risk assessing agent has to determine the FailureLevel of the risk assessed agent in a time slot after the time spot of his interaction i.e. in the post-interaction start time phase.

Case 5.2.1: If the time spot and the duration of the interaction (post-interaction start time phase) is limited to the time slot in which the risk assessing agent is at present as shown in Figure 5.8, then he can determine the FailureLevel (FFL) of the risk assessed agent by either considering his past-interaction history with the risk assessed agent (if any), or by soliciting recommendations from other agents, provided both have the same characteristics of the expectations of his future business interaction.

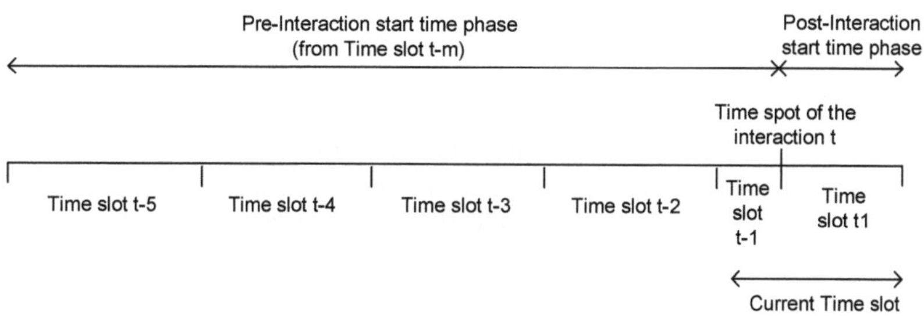

Fig. 5.8 The time spot and post-interaction phase of the interaction limited to the current period of time

Further explanation about the specific characteristics of the business activity is in Section 5.6.3.

Case 5.2.2: If the time spot or duration of the interaction (post-interaction start time phase) begins or extends to a future point in time from the current time slot in which the risk assessing agent is at present as shown in Figure 5.7, then he should utilize the determined FailureLevel of the risk assessed agent from the beginning of the time space to the current time slot to predict and determine the future FailureLevel (FFL) of the risk assessed agent in each of the post-interaction start time slots. A detailed explanation of how to determine the FailureLevel of the risk assessed agent in future time slots is given in Section 5.12.

5.6.3 *Considering the Specific Characteristics of the Business Activity While Performance Risk Analysis*

To consider the specific characteristics of his business association, the risk assessing agent can consider either his past interaction history with the risk assessed agent only if they:

a) are in the same time slot,
b) are with the same expectations, and
c) have the same significance value attached to each assessment criterion for his future interaction.

If this is the case, then the risk assessing agent can utilize the FailureLevel (AFL) that he had determined for the risk assessed agent in his past interaction (during the risk recording stage) as his FailureLevel (FFL) in the current interaction. This is based on the assumption made by Chang et al. [3] who state that the behavior of the risk assessed agent remains the same in a time slot, and subsequently the risk assessing agent can utilize the FailureLevel of the risk assessed agent from his past interaction (if it has the same expectations, significance and time slot of his future interaction) as his FailureLevel (FFL) in that time slot.

However, if the risk assessing agent does not have a past interaction history with the risk assessed agent in the expectations and in the time slot of his future interaction, or has a past interaction history in the part expectations of his future interaction, then in such cases the risk assessing agent can solicit recommendations about the risk assessed agent, from other agents for that particular time slot and for the assessment criterion or criteria of interest related to his expectations, and then aggregate them along with his past-interaction history (if any in the part expectations) to determine the FailureLevel (FFL) of the risk assessed agent in the post-interaction start time phase. A detailed explanation of how to determine the FailureLevel of the risk assessed agent in a time slot by either using direct past-interaction history and/or by soliciting recommendation from other agents is given in Section 5.9.

It may be the case that the risk assessing agent may neither have any past interaction history nor obtain any recommendations from other agents for the risk assessed agent against all the assessment criteria of his expectations in the current time slot of his interaction. In such cases, the risk assessing agent should determine the FailureLevel (FFL) of the risk assessed agent in the current time slot by using the approach proposed in case 5.2.2.

In each scenario discussed above, there are two important factors to be considered by the risk assessing agent. These are:

(a) the relative freshness of the risk assessed agent's capability in the time space of his association with him.
(b) the recommendations or opinions of other agents according to their credibility.

In the next sections, we discuss these factors further and present an approach that takes them into consideration during transactional risk analysis of a business activity.

5.7 Time Delay Factor to Consider the Freshness Status of the Risk Assessed Agent

As mentioned in Section 3.3.3, transactional risk is dynamic and it varies according to time. So it is very important during transactional risk analysis to consider how recent are the past interaction history or recommendations that are being considered with respect to the time spot of the current business activity.

For example let us consider that risk assessing agent 'A' in his current business association with agent 'B' forms the time space such that his previous interaction with agent 'B' falls in one of the pre-interaction start time slots. During performance risk analysis of that time slot, even though agent 'A' can consider his past interaction history to ascertain the FailureLevel of agent 'B' in that time slot (as discussed in scenario 5.1), he should not consider the exact Actual FailureLevel (AFL) value that he had assigned to the risk assessed agent during that time period as his PFL in that time slot. The reason for this may be that, since the time of the last interaction, the capability of agent 'B' to act according to the expectations may have declined as compared to the impression that agent 'A' had from the previous interaction. A similar approach is taken when using recommendations from other agents as shown in Figure 5.9.

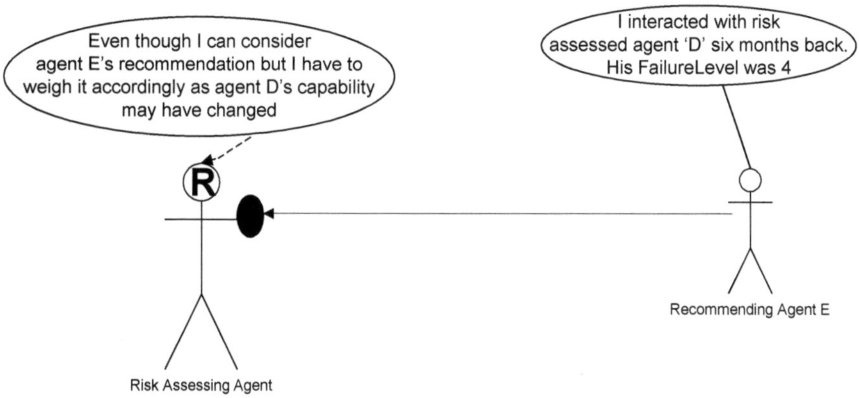

Fig. 5.9 Adjusting the recommendations to consider the freshness of the risk assessed agent's capability

So in order to avoid the effects of these scenarios, it is important for the risk assessing agent to adjust the FailureLevel value of the risk assessed agent accounting for the time delay factor. Also, considering the recency status of the risk assessed agent's capability allows the risk assessing agent to avoid modeling his behavior in the future, given that it may no longer be correct or current in terms of the expectations of his future interaction. This is particularly important when ascertaining the FailureLevel values of the risk assessed agent at a future point in time. Hence, the adjustment factor should place more weight on the recent FailureLevel values of the risk assessed agent rather than on his FailureLevel values in the older time slots, progressively reducing the effect of the older FailureLevel values in determining his FailureLevel value/s for a future point in time.

In Section 4.5, we mentioned that when the risk assessing agent has to determine the transactional risk for a future period of time, it formulates the time space such that the pre-interaction start time phase can adequately capture the previous impression of the risk assessed agent on which his performance risk in the post-interaction start time phase can be modeled. In this book, we consider that the risk assessing agent in such scenarios considers the FailureLevel of the risk assessed agent in at least 15 timeslots prior to the time spot of the interaction. This allows the risk assessing agent to capture the different variations in the risk assessed agent's behavior and inability to commit to the expectations, when utilizing it to determine his FailureLevel in future time slots.

To capture the dynamic nature of transactional risk and the freshness of the status of the risk assessed agent, we consider that the risk assessing agent from those pre-interaction start time slots should give more importance to the FailureLevel of the risk assessed agent in the first five time slots previous to the time spot of his interaction (t-1 till t-5 as shown in Figure 5.8) compared to the other time slots. For the importance to be given to the status or FailureLevel of the risk assessed agent in the other time slots of the pre-interaction start time phase (t-6 till t-15), the weight should be adjusted with a progressively declining value.

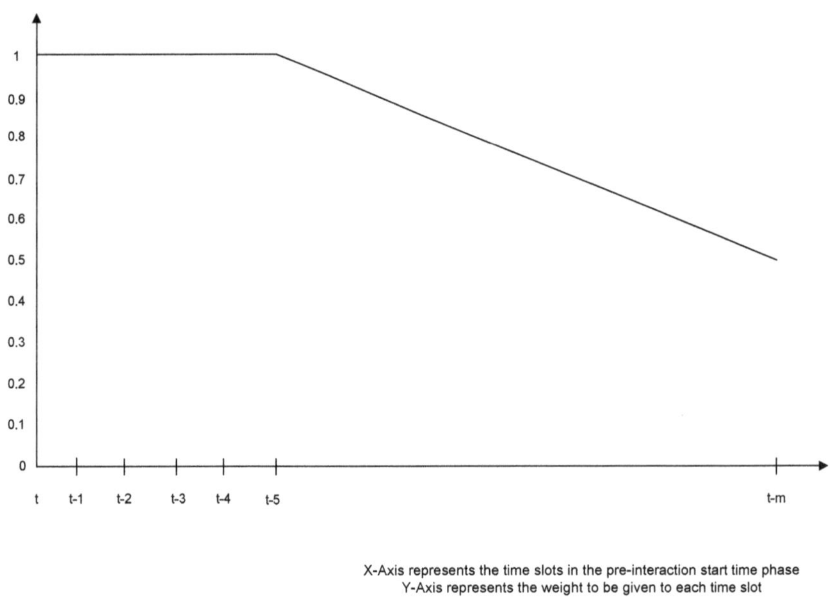

Fig. 5.10 The weight given to the risk assessing agent's capability in each time slot of the pre-interaction start time phase

We represent the weight to be given to the status of the risk assessed agent in a time slot before the time spot of the interaction by the variable '*w*' (represented in Figure 5.10) and is determined by:

$$w = \begin{cases} 1 & \text{if } m \le \Delta t \\ e^{\frac{-((n+\Delta t)-m)}{N}} & \text{if } m > \Delta t \end{cases} \quad (5.1)$$

where:

'w' is the weight or the time delaying factor to be given to the status of the risk assessed agent,
'n' represents the current time slot,
'm' represents the time slot for which the weight of adjustment is determined,
'Δt' represents the time slots from the time spot of the interaction in which the risk assessing agent will give more importance to the freshness of the status of the risk assessed agent,
'N' is the term which characterizes the rate of decay.

In the next section, we discuss the credibility of recommendations and their impact on determining the performance risk in the business activity.

5.8 Considering the Recommendations according to Their Credibility

As discussed in Section 3.5, when the risk assessing agent does not have his own past interaction history for some or all of the total assessment criteria of the expectations, then he solicits recommendations from other agents to initiate the process of risk assessment. The agents who return their recommendations are known as the recommending agents and they communicate their risk relationship to the risk assessing agent. But the risk assessing agent, before considering and assimilating their recommendation, should first determine whether or not the information that the recommending agent communicates is reliable and merits further consideration. This can be explained by taking an example. Let us consider the scenario of agent 'A', analyzing the perceived transactional risk in forming a business association with agent 'B' for a period of 1 year. Agent 'A' does not have a direct past interaction history with agent 'B' and, to facilitate the process of risk analysis, he solicits recommendations from other agents. Let us consider that he receives the recommendations from agents D, E, F and G as shown in Figure 5.11.

Let us consider that agents F and G may be related in some way to agent 'B' and to facilitate the process of agent 'B' in forming a business association with agent 'A', agents F and G are giving incorrect or inflated recommendations to agent 'A'. Considering such recommendations will result in agent 'A' calculating an incorrect performance risk value for agent 'B' that will in turn propagate when determining the perceived transactional risk in the business activity. Therefore, it is important to prevent or avoid such poor recommendations. To achieve this, we propose that the risk assessing agent divide the recommendations that he receives into two broad categories according to the source from which he receives them. Those categories are:

a) Unknown agents and
b) Known agents.

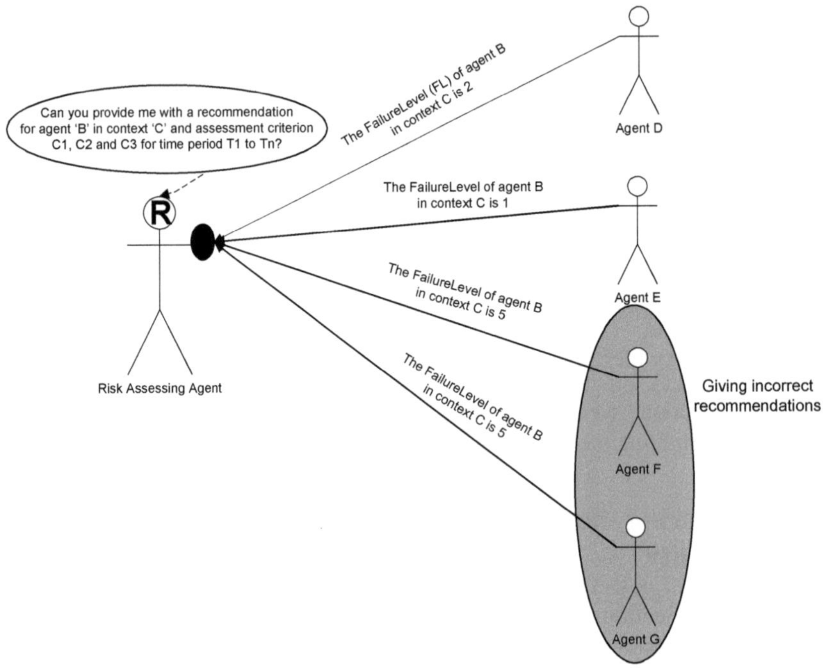

Fig. 5.11 Agents reply back to the risk assessing agent's query

We discuss these two categories of recommendations further in the next subsection.

5.8.1 Unknown Agents

Unknown agents are those with whom the risk assessing agent does not have previous experience in considering recommendations in the particular context.

5.8.2 Known Agents

Known agents are those with whom the risk assessing agent has had previous experience in soliciting and considering recommendations in a particular context. An agent known to the risk assessing agent implies that the risk assessing agent knows the level of reliability of that recommendation from that agent. These agents are further divided into two categories. These are:

5.8.2.1 Trustworthy Recommending Agents

Trustworthy agents are those whom the risk assessing agent believes and relies on to give correct opinions and recommendations.

5.8.2.2 Untrustworthy or Malicious Recommending Agents

Untrustworthy or malicious agents are those whom the risk assessing agent does not believe will provide correct recommendations. These have to be ignored by the risk assessing agent when he aggregates all recommendations to ascertain the performance risk of the risk assessed agent.

5.8.3 Recommending Agent's Credibility Value (RCV)

The process of classifying a recommending agent as either trustworthy or untrustworthy depends on the risk assessing agent's past experience with him. An important point to note here is that each recommending agent begins as an unknown agent to a risk assessing agent in communicating recommendations. Then, based on his experience and the correctness of the recommendations that he communicates, he is then classified into the relevant category of known agents.

To classify, determine and represent the reliability of a recommending agent when communicating recommendations, we propose the term 'Recommending Agent's Credibility Value' (*RCV*). RCV of a recommending agent is context-based, and we consider that the risk assessing agent maintains the RCV for a recommending agent in each context for which he took his recommendation. This value is used to determine whether or not the particular recommending agent is credible when giving recommendations in a particular context.

5.8.3.1 Scale to Represent the RCV of a Recommending Agent

As with Failure Scale, it is important to have a scale on which the RCV of a recommending agent in a context is presented for the risk assessing agent to understand in clear and unambiguous terms the level of credibility of the recommending agent in communicating recommendations. When a recommending agent communicates a recommendation, the risk assessing agent may find that one of two possible scenarios below may occur:

(a) The communicated recommendation by the recommending agent is exactly as determined by the risk assessing agent after the business activity,
(b) The communicated recommendation by the recommending agent varies either in a positive or negative way as determined by the risk assessing agent after the business activity.

So the scale to represent the RCV of a recommending agent should capture such levels of variation. In our approach, we consider the scale that ranges from (-5, 5) on a continuous scale for representing the credibility values of the recommending agents. We classify the range on this scale to differentiate between trustworthy and untrustworthy recommending agents.

5.8.3.2 Classification of a Recommending Agent according to His RCV

Trustworthy recommending agents are those agents whose RCV is within the specified range which is considered to be trustworthy by the risk assessing agent; whereas, untrustworthy agents are those agents whose RCV is beyond the specified range which is considered as trustworthy. In our approach, we consider that an agent whose RCV is within the range of (-1, 1) is considered as a known trustworthy recommending agent by the risk assessing agent. Within that range, a value of 0 specifies that the recommendation communicated by the recommending agent for the risk assessed agent is exactly similar to what the risk assessing agent finds after his interaction with that agent. A positive value to the range of 1 specifies that the risk assessing agent finds that the recommending agent recommends a lesser value for the risk assessed agent compared to what he determines for the risk assessed agent after his interaction. A negative value to the range of -1 specifies that the risk assessing agent finds that the recommending agent recommends a higher value for the risk assessed agent compared to what he determines for the risk assessed agent after the interaction. If the RCV of a recommending agent is beyond the range of (-1, 1), then it represents a big deviation from what the risk assessing agent finds after the interaction; and those agents are considered as untrustworthy agents. Further explanation of the way to determine the RCV of a recommending agent is given in Chapter 10.

5.8.3.3 Recommendation Table

As with transactional risk, it is important to note that the credibility of a recommending agent while giving recommendations is, also, context-dependent. For example, in the interaction scenario discussed earlier in this section, if agent 'D' is reliable and is a trustworthy recommending agent for agent 'A' in the particular context, then it does not mean that he is also a trustworthy recommending agent in other contexts as shown in Figure 5.12.

For these different contexts, agent 'A' should consider agent 'D's reliability in communicating recommendations in those contexts; and it is quite possible that in one context agent 'A' might consider agent 'D' as untrustworthy or unknown and hence has to consider D's recommendations accordingly. To consider scenarios like these, we consider that each risk assessing agent maintains his recommendation table (as shown in Table 5.1) in which the credibility values of each recommending agent along with the context it is recorded in.

Further while dealing with the recommendations we consider that:

- The risk assessing agent considers the recommendations from unknown and known trustworthy recommending agents and ignores the ones from untrustworthy or malicious agents as they mislead in determining the correct performance risk.
- Although the recommendations from unknown agents may be considered, more weight is given to the recommendations of known trustworthy agents as compared to the ones from unknown agents while assimilating them. One of the reasons for this is that the risk assessing agent is unsure whether the

unknown agent is a trustworthy recommending agent or a malicious recommending agent. So by giving more weight to the recommendations from known agents as compared to unknown agents, the risk assessing agent avoids the scenario of considering possible malicious recommendations and known trustworthy recommendations in the same way.

- Once the risk assessing agent considers the recommendation from a recommending agent then his credibility changes from unknown to either one of the type of known agents.

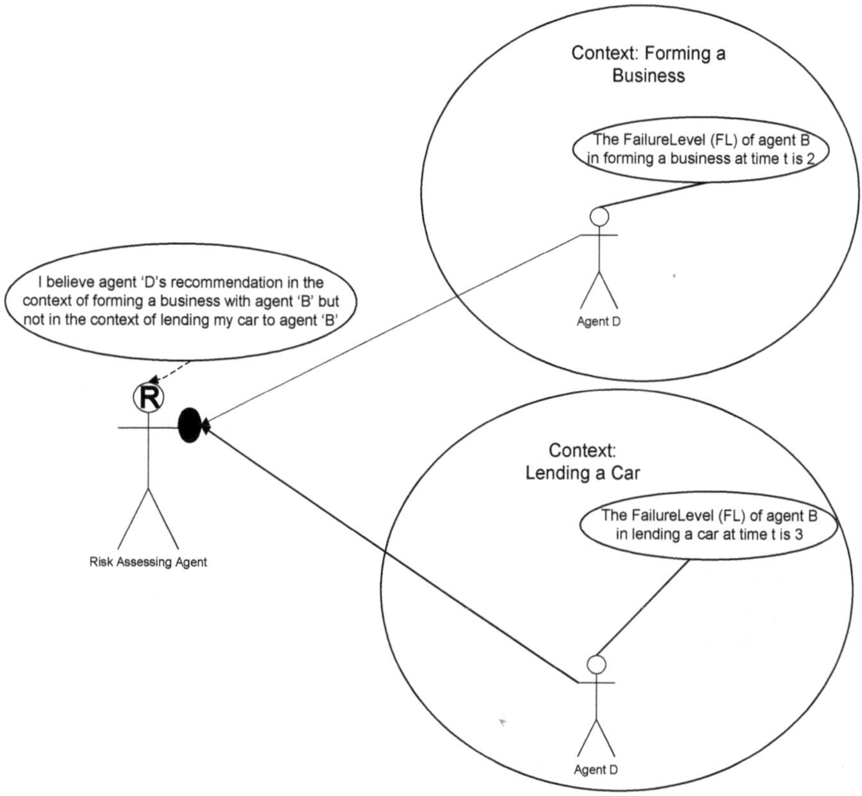

Fig. 5.12 Different recommendation scenarios

Table 5.1 Shows the recommendation table of agent 'A' in context C

Agent A's Recommendation table in Context C				
Recommending Agent's Name	Agent's ID	Time Period	RCV	Variation
Agent C	101	Oct 2009	0.58	0.01
Agent D	102	Nov 2009	1.25	0.05
Agent E	103	Jan 2010	-0.59	0.6
Agent F	104		UNKNOWN	-

After the risk assessing agent classifies the recommendations according to their credibility, weights them with the time delay factor and ascertains the significance of the assessment criteria (as discussed in Section 4.4), then he can proceed to the next steps of ascertaining the performance risk of the risk assessed agent in the pre-interaction and post-interaction start time phase. In both the time phases, the risk assessing agent has to determine the risk assessed agent's FailureLevel (which shows the level of non-commitment) in a time slot. The technique best suited to achieving this is probability theory which we will utilize to ascertain the FailureLevel of the risk assessed agent. We present our approach to determine the performance risk of the risk assessed agent in the next sections.

5.9 Determining the Probability of Failure in the Pre-interaction Start Time Slots

As mentioned in Scenario 5.1 the risk assessing agent can determine the FailureLevel of a risk assessed agent in a pre-interaction start time slot by

(a) using his own past interaction history, or
(b) soliciting recommendations from other agents.

We discuss these further in the next sub-sections.

5.9.1 Determining Direct Past Interaction-Based Probability of Failure in an Interaction

The direct past interaction-based probability of failure method refers to the risk assessing agent determining the probability of failure or FailureLevel in interacting with the risk assessed agent in a time slot, based on his past interaction history with him in that particular time slot. Further, the past interaction of the risk assessing agent with the risk assessed agent should be strictly according to the expectations and the same significance is attached to each assessment criterion, as in his future interaction with him. This is necessary to take the dynamic nature of transactional risk into consideration which varies according to the variation of the context and assessment criteria. Three scenarios arise when the risk assessing agent determines the FailureLevel of the risk assessed agent in a pre-interaction start time slot by considering his past interaction history with him in that time slot. They are:

Scenario 5.3: The assessment criteria of the risk assessing agent's previous interaction and their significance are the same as those of the expectations for his future interaction.

If the context, assessment criteria and their significance in the risk assessing agent's previous interaction with the risk assessed agent in a time slot of the pre-interaction start time phase are exactly the same as the expectations of his future interaction with him, then we propose that the risk assessing agent can utilize the risk relationship that he has formed with the risk assessed agent in that time slot, and consider the FailureLevel (AFL) that he had determined for the risk assessed agent in that interaction, as his FailureLevel (PFL) for that particular time slot. In Chapter 10, we explain in detail how the risk assessing agent ascertains the FailureLevel (AFL) of the risk assessed agent after his business association with him.

However as discussed in Section 5.7, in order to give more importance to the freshness of the status of the risk assessed agent which are in the time slots near or recent to the time spot of his interaction, the risk assessing agent should adjust the determined FailureLevel of the risk assessed agent in a pre-interaction start time slot ‘t-z’ (PFL_{Pt-z}), according to the weight that he wants to give to that time slot depending on where it falls in the time space of his current interaction. Hence, the FailureLevel (PFL) of the risk assessed agent ‘P’ in a pre-interaction start time slot ‘t-z’ based on the risk assessing agent's past interaction history with him in that time slot is represented by:

$$PFL_{Pt-z} = \text{ROUND}\ (w * AFL_{Pt-z}) \qquad (5.2)$$

where,

‘P’ represents the risk assessed agent,

‘t’ represents the time spot of the interaction,

‘z’ represents the number of time slots prior to the time spot of the risk assessing agent's interaction with the risk assessed agent,

‘w’ is the weight applied to the FailureLevel (AFL) of the risk assessed agent depending upon the time slot ‘t-z’.

The resultant value from equation 5.2 is rounded off to determine the crisp FailureLevel value for the risk assessed agent ‘P’ on the Failure Scale in the time slot ‘t-z’ (PFL_{Pt-z}).

For example, let us consider the logistics interaction scenario of agent ‘A’ with agent ‘B’ discussed in Section 4.3. Consider that agent ‘A’ has interacted with agent ‘B’ according to exactly the same assessment criteria mentioned in Figure 4.9; and in his previous interaction, he found the AFL of agent ‘B’ as 4 on the Failure Scale for time slot t_n. Considering that time slot t_n is t-3 (3rd time slot prior to the time spot of the interaction) then the PFL of agent ‘B’ in time slot ‘t-3’ of his current business interaction is:

$$PFL_{Bt-3} = \text{ROUND}\ (1 * 4) \rightarrow 4$$

Scenario 5.4: The assessment criteria of the risk assessing agent's previous interaction vary partly from the expectations of his future interaction, or the assessment criteria of the risk assessing agent's previous interaction are the same as those of his expectations, but the significance of these assessment criteria vary from those of the expectations for his future interaction.

Case 5.4.1: If the context of the risk assessing agent's previous interaction with the risk assessed agent in a time slot of the pre-interaction start time phase is the same, but the assessment criteria differ partly compared to the expectations of his future interaction, then we propose that:

(a) the risk assessing agent from his previous interaction should consider only those criteria which are similar to the assessment criteria for the expectations of his future interaction, and
(b) these should be utilized to determine the trustworthiness of the risk assessed agent in those, while considering the rest of the assessment criteria of his expectations by the reputation-based method (discussed in Section 5.9.2), and
(c) they should then be combined to determine the FailureLevel (PFL) of the risk assessed agent in that time slot (discussed in Section 5.11).

Case 5.4.2: If the assessment criteria of the risk assessing agent's previous interaction with the risk assessed agent in a time slot of the pre-interaction start time phase are identical to the expectations of his future interaction with it, but the significance of the criteria in his previous interaction vary from those of the assessment criteria of the expectations of the future interaction, then we propose that the risk assessing agent in such a case

(a) considers the assessment criteria from his previous interaction and
(b) utilize them to determine the trustworthiness of the risk assessed agent in these.

In both cases 5.4.1 and 5.4.2, the risk assessing agent cannot utilize the FailureLevel (AFL) that he had determined for the risk assessed agent in his previous interaction as the FailureLevel (PFL) of the risk assessed agent in the pre-interaction start time slot of his current interaction, as was done in the previous scenario. This is because in the first case, the FailureLevel (AFL) of the risk assessed agent determined in the past interaction is not exactly according to the expectations of his future interaction; and in the second case, the FailureLevel (AFL) of the risk assessed agent determined in the past interaction is not according to the significance of the expectations of his future interaction. Therefore in such cases, we propose that the risk assessing agent

(a) takes into consideration the relative 'assessment criteria' from his past interaction which are in the expectations of his future interaction, along with their corresponding 'Commitment Level' value that he had determined in his past interaction, and
(b) utilizes them to determine the risk assessed agent's trustworthiness in those assessment criteria according to the weight to be given to his status in that time slot.

An explanation about the commitment level of an assessment criterion was given in Section 3.6.3.

The trustworthiness of the risk assessing agent 'P' in an assessment criterion (C_n) and time slot 't-z' of the pre-interaction start time phase is determined by:

$$T_{PCn\ t\text{-}z} = (w * \text{Commitment Level}_{Cn}) \tag{5.3}$$

where:

'P' represents the risk assessed agent,

'Cn' represents the assessment criterion, in which the trustworthiness of the risk assessed agent 'P' is being determined,

'Commitment Level $_{Cn}$' represents the level of commitment of the risk assessed agent in assessment criterion 'Cn',

'*w*' is the weight applied to the commitment level of the risk assessed agent to consider his status in the time slot 't-z'.

If there is more than one assessment criteria in the risk assessing agent's past interaction history with the risk assessed agent which matches the expectations of his future interaction with him, then the risk assessing agent by using equation 5.3 should determine the trustworthiness of the risk assessed agent for each of those assessment criteria.

For example, let us consider that agent 'A' had a past interaction history with agent 'B' in assessment criteria C1 and C2 of Figure 4.9 in time slot t_{n-3}. Let us consider that he determined the commitment level of agent 'B' in those criteria are 1 and 0 respectively. Utilizing equation 5.3 the trustworthiness of agent 'B' in pre-interaction start time slot 't-3' is:

$$T_{BC1\ t\text{-}3} = (1 * 1) \rightarrow 1$$

$$T_{BC2\ t\text{-}3} = (1 * 0) \rightarrow 0$$

To consider the other assessment criteria of Figure 4.9 where the risk assessing agent does not have any past interaction history with the risk assessed agent, we propose that he solicit recommendations from other agents and utilize them to determine the reputation of the risk assessed agent in those. It should then:

a) utilize the trustworthiness and/or reputation value of the risk assessed agent determined in each assessment criterion of his expectations to ascertain his FailureLevel for each of them, and
b) combine the determined FailureLevel of each assessment criteria according to the significance in order to ascertain the FailureLevel (PFL) of the risk assessed agent in that time slot.

The methodology for the risk assessing agent to ascertain the FailureLevel of the risk assessed agent in a time slot by utilizing his trustworthiness (determined by using his past interaction history) and/or his reputation (determined from the recommendations from other agents) in the assessment criteria of the expectations is mentioned in Section 5.11.

Scenario 5.5: The assessment criteria of the risk assessing agent's previous interaction are completely different from the expectations of his future interaction.

If the context of the risk assessing agent's previous interaction with the risk assessed agent in a time slot of the pre-interaction start time phase is the same, but the assessment criteria are completely different from those of the future interaction, then the risk assessing agent cannot consider his past interaction history (AFL) as the FailureLevel (PFL) of the risk assessed agent of that time slot. In such cases, we propose that the risk assessing agent determine the FailureLevel of the risk assessed agent by utilizing the reputation-based probability of failure method. Further explanation about this method is given in the next sub-section.

5.9.2 Determining Reputation-Based Probability of Failure in an Interaction

The key steps in the reputation-based probability of failure method to determine the FailureLevel of a risk assessed agent are:

a) The risk assessing agent issues a reputation query to solicit recommendations from other agents by specifying the risk assessed agent, the particular assessment criterion or criteria and the time for which he wants the recommendations.
b) The agents who have had a previous interaction history with the risk assessed agent for the same time and the same assessment criterion or criteria, reply with the Risk Set as their recommendations.
c) The risk assessing agent classifies the recommendations according to their credibility and considers only the recommendations from trustworthy and unknown recommending agents in order to obtain an informed representation (as discussed in Section 5.8).
d) He then assimilates the recommendations to ascertain the reputation of the risk assessed agent according to the assessment criteria of his interests.

It is possible that the risk set with which the recommending agent communicates to the risk assessing agent may contain other assessment criteria apart from those which are of interest to him in his present interaction. Further, it is possible that the risk assessing agent may receive more than one recommendation for an assessment criterion of interest in a time slot from different agents. To consider all such types of recommendations, the risk assessing agent should determine the reputation of the risk assessed agent in the assessment criterion of interest, by assimilating all the recommendations that he can consider for that particular criterion from the recommending agents. The steps to be followed are:

a) Consider the 'Commitment Level' value for the particular assessment criterion of interest, from each trustworthy and unknown recommendation,
b) Adjust the commitment value according to the credibility of the recommending agent, and

c) Adjust it with the time decay weight factor to be given according to the distance from the time spot of the interaction.

The reputation of a risk assessed agent ‘P’ in an assessment criterion ‘Cn’ (Rep_{PCn}) in a pre-interaction time slot ‘t-z’ can be determined by assimilating the trustworthy and unknown recommendations by:

$$Rep_{PCn\ t\text{-}z} = (\alpha * (w * \frac{1}{K} (\sum_{i=1}^{K} RCV_i \oplus \text{Commitment Level}\ ^{i}_{Cn}))) + (\beta * (w * \frac{1}{J} (\sum_{o=1}^{J} \text{Commitment Level}\ ^{o}_{Cn}))) \quad (5.4)$$

where:

‘RCV_i’ is the credibility value of the trustworthy recommending agent ‘i’,
‘Commitment level $_{Cn}$’ is the level of commitment recommended by the recommending agent for assessment criterion ‘Cn’ for the risk assessed agent in the particular time slot ‘t-z’,
‘K’ is the number of trustworthy recommendations that the risk assessing agent gets for the risk assessed agent in assessment criterion ‘Cn’ in time slot ‘t-z’,
‘J’ is the number of unknown recommendations that the risk assessing agent gets for the risk assessed agent in assessment criterion ‘Cn’ in time slot ‘t-z’,
‘α and β’ are the variables attached to the parts of the equation which will give more weight to the recommendation from the trustworthy known recommending agents as compared to those from the unknown recommending agents. In general $\alpha > \beta$ and $\alpha + \beta = 1$,
‘*W*’ is the weight applied to consider the status of the risk assessed agent in time slot ‘t-z’.

As shown in equation 5.4, the reputation value of the risk assessed agent ‘P’ for assessment criterion ‘Cn’ is determined in two parts. The first part calculates the reputation value of the risk assessed agent ‘P’ in the assessment criterion ‘Cn’ by taking the recommendations of the trustworthy recommending agents. The second part calculates the reputation value of the same risk assessed agent ‘P’ in the same assessment criterion ‘Cn’ by taking the recommendations of the unknown recommending agents. The recommendations from the untrustworthy recommending agents are left out and not considered. The RCV of the trustworthy recommending agent is also considered while assimilating his recommendation by using the adjustment operator ‘$\oplus$’. This takes into consideration the accurate recommendation from the trustworthy recommending agent according to the credibility and accuracy by which he communicates his recommendations. The rules for the adjustment operator ‘$\oplus$’ are:

$$a \oplus b = \begin{cases} a + b, & \text{if } 0 \leq (a + b) \leq 1 \\ 1, & \text{if } (a + b) > 1 \\ 0, & \text{if } (a + b) < 0 \end{cases}$$

In order to give more importance to the recommendations from the trustworthy recommending agents compared to ones from the unknown recommending agents, variables are attached to the two parts of the equation. These variables are represented by α and β respectively. It depends upon the risk assessing agent how much weight he wants to assign to each type of recommendation. Furthermore, each recommendation for the risk assessed agent in a time slot is adjusted according to the weight to be given to the status of the risk assessed agent in that time slot. The risk assessing agent should utilize equation 5.4 to determine the reputation of the risk assessed agent either in all or in part of the assessment criteria, in a pre-interaction start time slot for which he does not have any past interaction history.

To explain with an example, let us consider that the risk assessing agent 'A' solicits recommendations for agent 'B' in time slot 't-3' and in assessment criteria C3 and C4 of Figure 4.9. Let us assume that he receives replies from agents 'E', 'F', 'G' and 'H' as shown below:

{Agent 'E', Agent 'B', Context C, 3, ((C3, 1), C5, 0)), 15000, t-3, t-3}

{Agent 'F', Agent 'B', Context C, 4, ((C3, 0)), 10000, t-3, t-3}

{Agent 'G', Agent 'B', Context C, 5, ((C3, 1), (C4, 1)), 5000, t-3, t-3}

{Agent 'H', Agent 'B', Context C, 4, ((C4, 0), (C10, 1), 16000, t-3, t-3}

As can be seen, the recommendations are in different assessment criteria which are part of what the risk assessing agent 'A' needs for his assessment. The next step for agent 'A' is to classify the recommendations according to their credibility. Let us consider that in context 'C' agent 'A' classifies agent 'E' and agent 'H' as trustworthy recommending agent with RCV of 0.15 and -0.69 respectively and agent 'F' and agent 'G' as unknown recommending agents. So utilizing equation 5.4 the reputation of agent 'B' in assessment criteria C3 and C4 are:

$$\text{Rep}_{\text{BC3 t-3}} = (0.6 * (1 * (0.15 \oplus 1))) + (0.4 * (1 * \frac{1}{2} (0+1))) \rightarrow 0.8$$

$$\text{Rep}_{\text{BC4 t-3}} = (0.6 * (1 * (-0.69 \oplus 0))) + (0.4 * (1 * (1))) \rightarrow 0.4$$

By using the approach discussed, the risk assessing agent can determine the reputation of the risk assessed agent for a particular assessment criterion in a time slot. However, it may be possible that:

a) the risk assessing agent may not receive any recommendation for the risk assessed agent 'P' in an assessment criterion 'C_n' of his interest from the expectations in a time slot.
b) the recommendation in an assessment criterion might not be interpreted or recognized in a similar way by the risk assessing agent.

In this case, to conduct a sensible risk analysis, a value of 0 can be assigned as the reputation of the risk assessed agent for that assessment criterion 'C_n' ($Rep_{PCn\ t-z}$) in that time slot. However, such outcomes might not be fair to the risk assessed agent as he could still meet the assessed criteria even though the risk assessing agent has no prior knowledge of these. Furthermore, assigning an incorrect reputation value to the risk assessed agent at this stage will be propagated further as:

a) The FailureLevel of risk assessed agent determined in a pre-interaction start time slot is dependent on the FailureLevel for each assessment criterion in that time slot.
b) The performance risk of the risk assessed agent in the post-interaction start time phase is dependent on the FL values in the pre-interaction start time phase.
c) The sub-category of Financial Risk is dependent on the performance risk determined in the post-interaction start time phase, and
d) The level of transactional risk is dependent on the performance risk and financial risk determined in the post-interaction start time phase.

So, in the absence of information being present for an assessment criterion, it is important to make an informed decision about the reputation of a risk assessed agent in that assessment criterion. In this book, we address this problem by proposing an ontology-based semantic similarity model where, in the absence of both direct past interaction history and recommendations from other agents for an assessment criterion, the reputation value of a risk assessed agent in that assessment criterion can be determined by utilizing the level of similarity with the other criteria present. We explain this further in the next section.

5.10 Ontology-Based Criteria Similarity Matching Framework

Traditional semantic similarity models focus on measuring semantic similarity between nodes in semantic networks. Semantic networks refer to the graphic notations comprising arcs and nodes, in which nodes represent concepts and arcs represent relations between concepts [4]. Semantic networks can be used to represent simple knowledge in specific domains and a typical example is WordNet. However, limitations of semantic networks include:

1) nodes are usually single words and cannot be defined by properties; and
2) arcs cannot be defined by restrictions and characteristics [5].

Compared with semantic networks, ontologies are a form of knowledge representation with more complex attributes. Ontologies consist of concepts and relations between concepts [6]. The advantages of ontologies include:

1) concepts can be defined by both datatype and object properties (relations);
2) object properties (relations) can be defined by multiple restrictions and characteristics.

In terms of the comparison, it is not difficult to observe that ontologies can be employed to represent knowledge with more complex structures. Meanwhile, with the emergence of ontologies, new forms of semantic similarity models were developed in order to measure concept similarity in the ontology environment, known as ontology-based semantic similarity models [7]. We utilize such characteristics to develop an ontology-based assessment criteria similarity matching framework for performance risk assessment.

5.10.1 Ontology-Based Assessment Criteria Similarity Matching Framework for Performance Risk Assessment

To have an ontology based assessment criteria similarity matching model, we propose that the risk assessing agent will store the assessment criteria of his previous interactions with the risk assessed agent in a database repository as shown in Figure 5.13 [8]. However, this is not sufficient and may again lead to the scenarios of mis-interpretation discussed in the previous case where the risk assessing agent may assume that there are no replies from the recommending agents for those assessment criteria. To avoid such situations, we propose the use of a knowledge base which stores generic ontologies representing relationships between context-specific risk assessment criteria as shown in Figure 5.13.

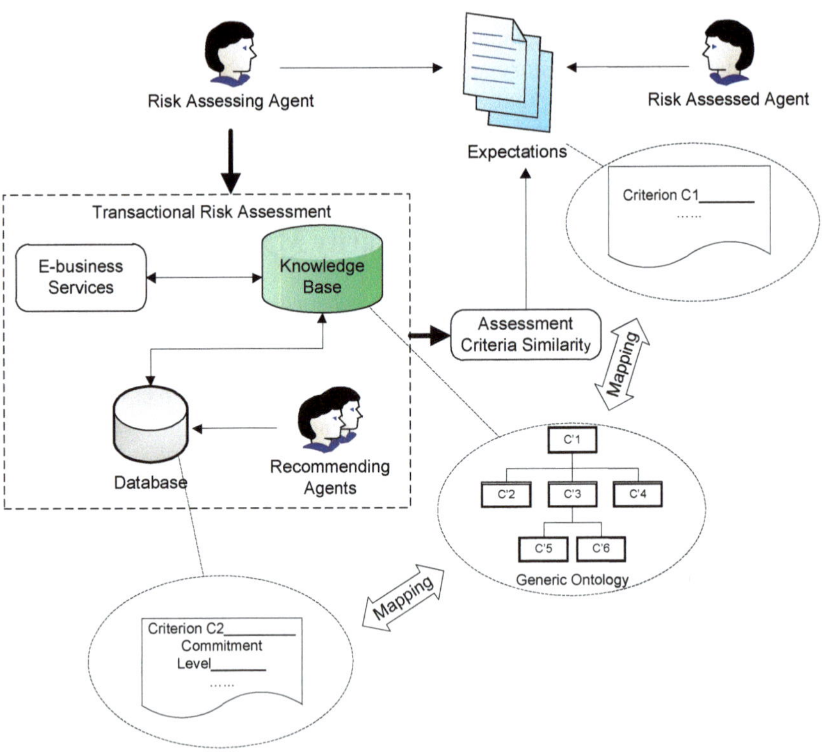

Fig. 5.13 Criteria Similarity Measure Framework for Performance Risk Assessment

Furthermore, each generic ontology provides a shared representation of context specific concepts and is regarded as a hierarchical structure of concepts linked by is-a relations. By using such a structure, an assessment criterion can be annotated by one or more concepts from a relevant generic ontology and a semantic similarity model can be used to measure the similarity between the current assessment criteria and the historical assessment criteria. In this book, we develop a similarity model that is designed based on the theory of Rada et al. [9] distance-based metric as shown in Table 5.2. This approach calculates the semantic distance between two concepts in terms of the shortest distance in a semantic network.

The scope of the similarity value between any two concepts is within the interval [0, 1], where 0 stands for nothing similar and 1 stands for completely similar. It needs to be noted that each ontology should represent an assessment dimension in a disjoint context of services. Therefore, this semantic similarity model cannot measure the similarity between two assessment criteria in different contexts, so the similarity value should be 0 in that case. For example, the similarity value between the criterion of latency in a video service and that of an audio service should be 0, since there is no direct relationship between the two service contexts and a service provider's performance in the audio service cannot affect his/her performance in the video service.

In order to explain the proposed semantic similarity model, we provide a case study to describe the application of this model in the domain of cloud computing. We make use of a real-use case scenario adopted from Amazon Web Services™ (http://aws.amazon.com) and consider that a consumer (risk assessing agent) in Virginia wants to use an Amazon Elastic Compute Cloud (EC2) service (http://aws.amazon.com/ec2/). In the expectations, the risk assessing agent wants to use a small instance of the EC2 service (Windows) and intends to obtain the performance risk of the small instance for the price criterion. However, the risk assessing agent has not had any previous business associations with Amazon on the usage of the small instance but has a transaction history with Amazon on the usage of the large instance (Windows) and the usage of the high CPU medium instance (Windows). To utilize that past interaction history in the current case, the price ontology in the context of the Amazon EC2 services is defined as shown in Figure 5.14.

Using the concepts in the defined ontology and the proposed semantic similarity model, the similarity between the price for the small instance (S) and the price for the large instance (L) can be obtained as shown below

$$sim(S,L)=0.667$$

Table 5.2 Algorithm to determine the semantic distance between two assessment criteria

Input: c1, c2 are two assessment criteria, O is an e-business ontology which consists of concepts (c'1...c'n) linked by *is-a* relations and its maximum depth is d.
Output: sim(c1, c2) – the similarity between c1 and c2.
Algorithm:

```
begin
for i = 1 to n
        if c'[i] ∈ c1 then
                Put c'[i] into an array C1;
        else if c'[i] ∈ c2 then
                Put c'[i] into an array C2;
        end if
end for
k = count (C1);
l = count (C2);
for i = 1 to k
        s = 2d;
        for j = 1 to l
                A[i][j] = the shortest distance between C1[i] and C2[j] in O;
                if A[i][j] < s then
                        s = A[i][j];
                end if
        end for
        d' = d' + s;
end for
for j = 1 to l
        t = 2d;
        for i = 1 to k
                if A[i][j] < t then
                        t = A[i][j];
                end if
        end for
        d' = d' + t;
end for
```

$$sim(c1,c2) = 1 - \frac{d'}{(k+l)x2d'} \quad \mathrm{sim}(c1, c2) = 1 - \frac{d`}{(k+l)\times 2d'};$$

end

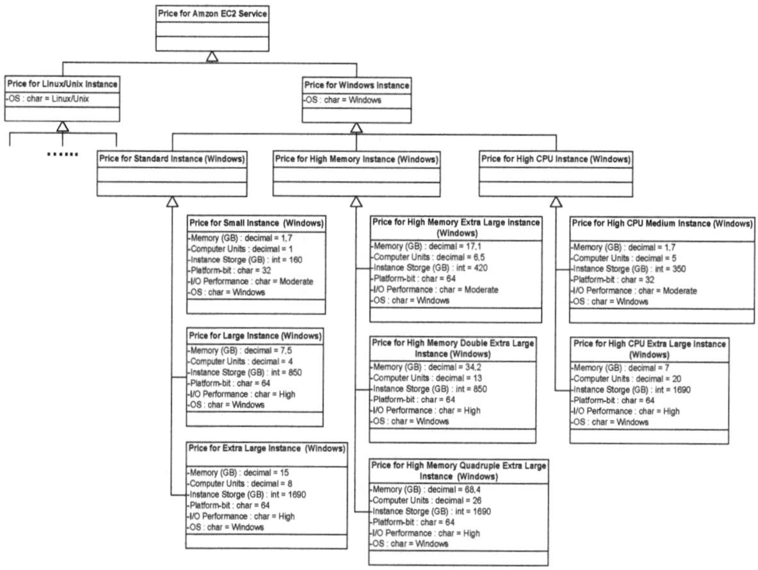

Fig. 5.14 Abbreviated view of a price ontology in the context of Amazon EC2 services

Subsequently, the similarity between the price for the small instance (S) and the price for the high CPU medium instance (CM) can be calculated by

$$sim(S,CM)=0.333$$

Once the framework has measured the similarity between the current assessment criterion and the historical assessment criterion, the degree of similarities are used as weights to determine the level of commitment of the risk assessed agent to the assessment criterion of the expectations.

By using the above framework, the risk assessing agent can determine the trustworthiness or reputation of the risk assessed agent 'P' in a time slot 't-z' of the pre-interaction start time phase and in an assessment criterion (C_n) by considering the level of similarity with other criteria by:

$$T_{PCn\ t\text{-}z} = (sim(C_n \rightarrow C_z) * \text{Commitment Level}_{Cz} * w)$$

or

$$Rep_{PCn\ t\text{-}z} = (sim(C_n \rightarrow C_z) * \text{Commitment Level}_{Cz} * w) \quad (5.5)$$

where:

'P' represents the risk assessed agent,

'Cn' represents the assessment criterion, in which the trustworthiness or reputation of the risk assessed agent 'P' is being determined,

'Cz' represents the criterion with which the similarity of assessment criterion 'Cn' is being determined,

'Commitment Level $_{Cz}$' represents the level of commitment of the risk assessed agent in assessment criterion 'Cz',

sim($C_n \rightarrow C_z$) represents the level of similarity between assessment criteria C_n and C_z,

'*w*' is the weight applied to the commitment level of the risk assessed agent to consider his status in the time slot 't-z'

5.11 Ascertaining the FailureLevel (PFL) of the Risk Assessed Agent in a Pre-interaction Start Time Slot

Once the risk assessing agent ascertains

(a) the trustworthiness of the risk assessed agent for part or all of the assessment criteria of the expectations by using his past interaction history (discussed in scenario 5.2), and/or
(b) the reputation of the risk assessed agent by using recommendations from the other agents in those assessment criteria in which he does not have any past interaction history (discussed in Section 5.9.2), and
(c) in the absence of both by using the Ontology-based similarity matching framework (discussed in Section 5.10),

then he should combine them to determine the FailureLevel (PFL) of the risk assessed agent in the pre-interaction start time slot 't-z'. To achieve this, the risk assessing agent has to first ascertain the FailureLevel of the risk assessed agent for each assessment criterion of the expectations, from the determined trustworthiness or reputation in these.

The trustworthiness or the reputation of the risk assessed agent in an assessment criterion shows his level of capability to meet the particular criterion. To determine the FailureLevel of the risk assessed agent for that criterion, the extent of his inability to complete the given assessment criterion has to be determined. To achieve this, we propose that the:

(a) risk assessing agent should map the trustworthiness or the reputation of the risk assessed agent in each assessment criterion of the expectations in a pre-interaction start time slot 't-z', on the Failure Scale (FS). By doing so, the risk assessing agent knows the capability of the risk assessed agent to meet that assessment criterion on the Failure Scale, in the time slot 't-z'.
(b) He can then determine the probability of failure of the risk assessed agent in committing to that assessment criterion in that time slot according to the expectations, by ascertaining the difference between what he expects of that assessment criterion, and how far the risk assessed agent can fulfill it according to his trustworthiness or reputation for that criterion. The value achieved gives the probability of failure to meet that assessment criterion in that time slot.

(c) The FailureLevel of the assessment criterion in that time slot is then achieved by mapping the probability of failure of that assessment criterion to the Failure Scale (which is between 0 and 5).

5.11.1 Mapping the Trustworthiness or Reputation of the Risk Assessed Agent on the Failure Scale

As mentioned earlier, the levels on the Failure Scale between 0 and 5 represent varying severity and magnitudes of failure. Hence, to determine the FailureLevel of the risk assessed agent in an assessment criterion, his trustworthiness or reputation for that criterion should be mapped on the range of (0, 5) on the Failure Scale, as it is within these levels that his capability to complete the assessment criterion has to be ascertained on the Failure Scale. The trustworthiness or the reputation of the risk assessed agent in an assessment criterion can be represented on the Failure Scale (FS) by equation 5.6:

$$T_{PCn\ t\text{-}z\ FS} = \text{ROUND}\ (T_{PCn\ t\text{-}z} * 5)$$
or,
$$Rep_{PCn\ t\text{-}z\ FS} = \text{ROUND}\ (Rep_{PCn\ t\text{-}z} * 5) \qquad (5.6)$$

where:

‘$T_{PCn\ t\text{-}z\ FS}$’ represents the trustworthiness of the risk assessed agent in time slot ‘t-z’ and in assessment criterion ‘Cn’ on the Failure Scale,

‘$T_{PCn\ t\text{-}z}$’ represents the trustworthiness of the risk assessed agent in assessment criterion ‘Cn’ and in time slot ‘t-z’,

‘$Rep_{PCn\ t\text{-}z\ FS}$’ represents the reputation of the risk assessed agent in time slot ‘t-z’ and in assessment criterion ‘Cn’ on the Failure Scale,

‘$Rep_{PCn\ t\text{-}z}$’ represents the reputation of the risk assessed agent in assessment criterion ‘Cn’ and in time slot ‘t-z’.

5.11.2 Determining the Probability of Failure of the Risk Assessed Agent in Committing to an Assessment Criterion

The risk assessing agent expects the risk assessed agent to complete the assessment criterion according to his expectations. This expectation of the risk assessing agent can be quantified with a value of 5 on the Failure Scale, as it represents the lowest probability of failure of the assessment criterion and expresses the maximum commitment by the risk assessed agent to the expectations. The probability of failure to achieve an assessment criterion ‘Cn’ according to the expectations in interacting with the risk assessed agent ‘P’ in a time slot ‘t-z’, according to his trustworthiness or reputation in this can be determined by equation 5.7:

$$\text{Probability of Failure}_{\text{PCn t-z}} = \left(\frac{5 - T_{\text{PCn t - zFS}}}{5}\right) * 100$$

or,

$$\text{Probability of Failure}_{\text{PCn t-z}} = \left(\frac{5 - Rep_{\text{PCn t - zFS}}}{5}\right) * 100 \tag{5.7}$$

5.11.3 Mapping the Probability of Failure to Determine the FailureLevel of a Risk Assessed Agent

The risk assessing agent from the determined probability of failure in Section 5.11.2 can determine the FailureLevel (PFL) of the risk assessed agent 'P' in assessment criterion 'Cn' and in time slot 't-z' on the Failure Scale ($PFL_{\text{PCn t-z}}$) by equation 5.8:

$$PFL_{\text{PCn t-z}} = \text{LEVEL}\ (\text{Probability of Failure}_{\text{PCn t-z}}) \tag{5.8}$$

5.11.4 Determining the Crisp FailureLevel of a Risk Assessed Agent in a Pre-interaction Start Time Slot

By using the above steps, the risk assessing agent should determine the FailureLevel of the risk assessed agent for each assessment criterion of the expectations in a pre-interaction start time slot. Once that is done, then the crisp FailureLevel of the risk assessed agent 'P' in a pre-interaction start time slot 't-z' ($PFL_{\text{Pt-z}}$) is determined by weighing his FailureLevel to complete each assessment criterion of the expectations in that time slot, with the significance of the assessment criterion (discussed in Section 4.5) as shown in Equation 5.9.

$$PFL_{\text{Pt-z}} = \text{ROUND}\ \left(\sum_{n=1}^{y} S_{Cn} * PFL_{\text{PCn t-z}}\right) \tag{5.9}$$

where:

'S_{Cn}' is the significance of the assessment criterion 'Cn';
'$PFL_{\text{PCn t-z}}$' represents the FailureLevel of the risk assessed agent 'P' in assessment criterion 'C_n' in time slot 't-z'; and
'y' is the number of assessment criteria in the expectations.

To explain with an example, let us consider the trustworthiness (explained in case 5.4.2) and reputation (explained in Section 5.9.2) value determined for agent 'B' in assessment criteria C1-C4 are as follows:

$$T_{\text{BC1 t-3}} = 1;\ \ T_{\text{BC2 t-3}} = 0;\ \ Rep_{\text{BC3 t-3}} = 0.8;\ \ Rep_{\text{BC4 t-3}} = 0.4$$

Utilizing equation 5.6 to map the determined commitment level of agent 'B' on the Failure Scale, agent 'A' obtains the following values:

$$T_{BC1\ t\text{-}3\ FS} = 5;\ T_{BC2\ t\text{-}3\ FS} = 0;\ Rep_{BC3\ t\text{-}3\ FS} = 4;\ Rep_{BC4\ t\text{-}3\ FS} = 2$$

By using equation 5.7 and equation 5.8, the FailureLevel of agent 'B' (PFL) in time slot t-3 and in assessment criteria C1-C4 determined on the Failure Scale are:

$$PFL_{BC1\ t\text{-}3} = 5;\ PFL_{BC2\ t\text{-}3} = 0;\ PFL_{BC3\ t\text{-}3} = 4;\ PFL_{BC4\ t\text{-}3} = 2$$

The crisp FailureLevel of agent 'B' in pre-interaction start time slot 't-3' determined by equation 5.9 and by the significance values of the assessment criteria of Table 4.6 is:

$$PFL_{Bt\text{-}3} = \text{ROUND}\ (2.788) \rightarrow 3$$

By using the proposed methodology, the risk assessing agent should ascertain the FailureLevel (PFL) of the risk assessed agent in each time slot of the pre-interaction start time phase according to the expectations of his future interaction. He can then utilize that analysis to ascertain the FailureLevel (FFL) of the risk assessed agent in the time slots of the post-interaction start time phase. As the FailureLevel of the risk assessed agent in the pre-interaction start time slots is according to the expectations of his future interaction, the determined FailureLevel in the time slots of the post-interaction start time phase will also be strictly according to the expectations of the risk assessing agent's future interaction.

5.12 Ascertaining the FailureLevel (FFL) of the Risk Assessed Agent in a Post-interaction Start Time Slot

As discussed in Section 4.5, the risk assessing agent's actual period of interaction with the risk assessed agent is represented by the post-interaction start time phase of the time space, and this period of time ranges from the time spot of the interaction to the end of the time space. Two scenarios arise when the risk assessing agent determines the FailureLevel of the risk assessed agent in the post-interaction start time phase. They are:

Scenario 5.6: The post-interaction start time phase of the risk assessing agent's interaction with the risk assessed agent is limited to the current time slot in which he is at present.

If the time spot and the duration of the risk assessing agent's interaction with the risk assessed agent is limited to the current time slot (as shown in Figure 5.8), then the risk assessing agent can determine the FailureLevel (FFL) of the risk assessed agent in the post-interaction start time slot by:

Case 5.6.1: Using his past interaction history with the risk assessed agent, if it is in the same expectations and time slot of his future interaction.

If the risk assessing agent has a past interaction history with the risk assessed agent with the same characteristics as those of the expectations of his future interaction, then he can consider the risk relationship of his previous interaction with the risk assessed agent and utilize the FailureLevel (AFL) which he had ascertained for the risk assessed agent in that previous interaction, as his FailureLevel (FFL) in the post-interaction start time slot. This is based on the assumption mentioned earlier that the FailureLevel of the risk assessed agent in a time slot remains constant [3].

Case 5.6.2: Using a combination of his past interaction history and the recommendations from other agents.

If the risk assessing agent has a past interaction history with the risk assessed agent in the same context and in the same time slot of his future interaction, but in part of the assessment criteria of the expectations, then he should utilize that part of the assessment criteria and their corresponding 'Commitment Level' values to determine the trustworthiness of the risk assessed agent for those assessment criteria as discussed in scenario 5.4. He should then solicit recommendations from other agents for the remaining assessment criteria of the expectations by issuing a reputation query, and then assimilate them to ascertain the reputation of the risk assessed agent for those assessment criteria as discussed in Section 5.9.2.

However, in each of the cases discussed previously (scenario 5.4 and Section 5.9.2), the risk assessing agent adjusts the trustworthiness and/or the reputation of the risk assessed agent by the time decay variable '*w*' depending upon the time slot in the pre-interaction start time phase. In the present case, as he is in the current time slot the value for the variable '*w*' should be 1. Based on the determined trustworthiness or the reputation of the risk assessed agent for each assessment criterion of the expectations, the risk assessing agent can then determine the FailureLevel (FFL) of the risk assessed agent in the post-interaction start time slot by using the methodology proposed in Section 5.11.

It may be the case that the risk assessing agent does not have any past interaction history with the risk assessed agent in the time slot of his interaction, nor does he get recommendations from other agents for all the assessment criterion of the expectations in the time slot of his interaction. In this case, the risk assessing agent cannot utilize the above proposed methodology to determine the FailureLevel (FFL) of the risk assessed agent in the post-interaction start time phase of his interaction, and should utilize the methodology proposed in scenario 5.7 to determine the FailureLevel (FFL) of the risk assessed agent in that time phase.

Scenario 5.7: The post-interaction start time phase of the risk assessing agent's interaction with the risk assessed agent begins and extends to a point in time in the future.

As discussed in the earlier sections, if the time spot or the duration of the risk assessing agent's interaction with the risk assessed agent extends to a future point in time (as shown in Figure 5.7), then the risk assessing agent has to determine the FailureLevel (FFL) of the risk assessed agent in those time slots by utilizing the prediction methods based on the previous impression that he had about the risk assessed agent. Hence, in our method we propose that the risk assessing agent, in order to determine the future FailureLevel of the risk assessed agent at time slot 't1' of the post-interaction phase in Figure 5.7, should consider all his FailureLevel values from the beginning of the time space to the time slot preceding it, i.e. to time slot 't-1' and utilize it to predict the FailureLevel values in time slot t1. Two cases arise when the risk assessing agent has to ascertain the FailureLevel (FFL) of the risk assessed agent in the future period of time of his interaction.

Case 5.7.1: The determined FailureLevel (PFL) of the risk assessed agent in the pre-interaction start time slots has features of either stochastic variation or trends in variation as shown in Figure 5.15.

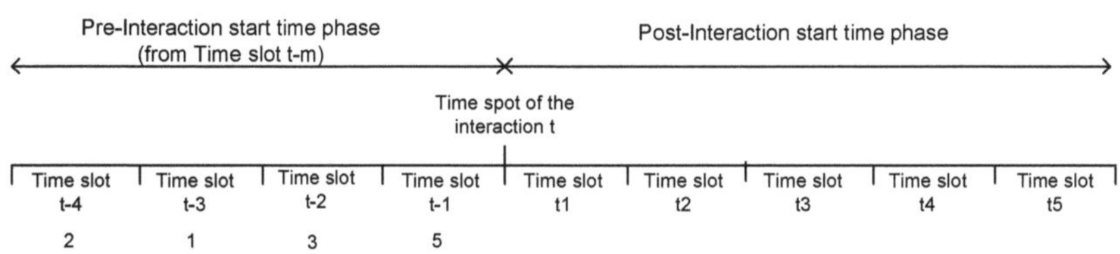

Fig. 5.15 Showing the stochastic variation or trends in the PFL of agent 'B'

In this case, we propose that the risk assessing agent while determining the FailureLevel (FFL) of a risk assessed agent in a time slot of the post-interaction start time phase, should determine the magnitude of occurrence of each level of failure within the domain of (0, 5) on the Failure Scale in that time slot, rather than determining a crisp FailureLevel as he does in the pre-interaction start time slots. This is because the performance risk of the risk assessed agent in the future period of time reflects uncertainty as it is a point in time in the future; and subsequently, when the FailureLevel series of the risk assessed agent contains variability in the pre-interaction start time slots, the uncertainty in his behavior at a future period of time should be captured, while ascertaining his FailureLevel during that time period. This uncertainty is not totally captured when it is being represented by a crisp FailureLevel value. Hence, to take this into consideration, we suggest that the risk assessing agent should ascertain the likelihood of the occurrence of each severity level of failure on the Failure Scale.

As the time space is chosen so that the number of time slots in the pre-interaction start time phase is at least 15, then we consider that, by utilizing that analysis, the determination of the FailureLevel (FFL) of the risk assessed agent in a post-interaction start time slot on the Failure Scale (FS) will follow a representation of Gaussian series. As discussed earlier, the domain of the Failure Scale ranges from (-1, 5), with -1 denoting 'Unknown' level of failure. So the FailureLevel (FFL) of a risk assessed agent in the post-interaction start time slot should be determined in the domain of (0, 5) on the Failure Scale. Within this domain, there are six possible levels of failure. To determine the risk assessed agent's FailureLevel (FFL) at time slot 't1' within each of those levels, let us suppose that the risk assessing agent has determined the FailureLevel of the risk assessed agent in each time slot from the beginning of the time space till time slot 't-1' as represented by:

$$\{FL_{t-K,}\ldots\ldots FL_{t-3}, FL_{t-2}, FL_{t-1}\}$$

where k is the number of time slots preceding the one in which the FFL is being determined.

The mean FailureLevel (μ_{FL}) is calculated as:

$$\mu_{FL} = \frac{1}{K} \sum_{i=1}^{K} FL_i \tag{5.10}$$

Accordingly, the unbiased Sample Variance (σ^2) is:

$$\sigma^2 = \frac{1}{K-1} \sum_{i=1}^{K} (FL_i - \mu_{FL})^2 \tag{5.11}$$

Since FFL ~ (μ, σ^2), then for any random variable FFL according to Gaussian distribution [10], the probability of FFL in a given range within the domain of (0, 5) on the Failure Scale can be determined according to equation 5.12.

$$P(a < FFL \leq b) = \frac{1}{\sqrt{2\Pi}\sigma} \int_{\frac{a-\mu}{\sigma}}^{\frac{b-\mu}{\sigma}} e^{\frac{-t2}{2}} dt \tag{5.12}$$

By using equation 5.12, the risk assessing agent can determine the probability of occurrence of the different severity of failure and their likelihood of occurrence in a particular time slot of his interaction in future. This also represents the variability in the behavior of the risk assessed agent over that particular future period of time of his interaction. As the PFL of the risk assessed agent is determined strictly according to the expectations of the business activity, the determined severities of failure too are strictly according to the expectations of interaction between the risk assessing agent and the risk assessed agent.

To explain with an example, let us continue the logistics interaction scenario between agent 'A' and agent 'B' mentioned in Section 4.3. Let us consider that agent 'A' has 15 time slots in the pre-interaction start time phase and he determines the FailureLevel (PFL) of agent 'B' in those time slots as:

$$\{1, 2, 3, 3, 3, 2, 4, 2, 3, 3, 1, 4, 5, 4, 3\}$$

Utilizing equations 5.10 - 5.12 the probability of the occurrence of each level of failure on the Failure Scale in time slot t1 of the time space as shown in Figure 5.15 is as shown in Figure 5.16.

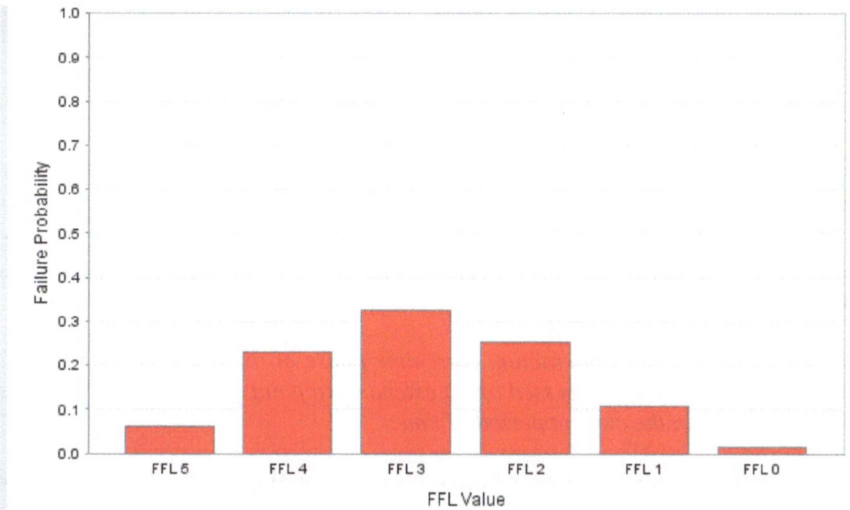

Fig. 5.16 Showing the probability of occurrence of different levels of Failure in time slot t1

If there is more than one time slot in the post-interaction start time phase of agent 'A's interaction with agent 'B' as shown in Figure 5.15, then the risk assessing agent has to determine the FailureLevel (FFL) of the risk assessed agent in each time slot of the post-interaction start time phase ('t1' till 't5'), to ascertain the performance risk in interacting with it.

To ascertain the FailureLevel (FFL) of the risk assessed agent in the post-interaction start time slot 't2', the risk assessing agent, after determining the likelihood of occurrence of each severity level of failure in interacting with the risk assessed agent in the post-interaction start time slot 't1', should take the FailureLevel with the highest probability of occurrence as the FailureLevel of the risk assessed agent in time slot 't1'. He should then consider the time slots from the beginning of the time space till time slot 't1' as shown in Figure 5.15, and utilize equations 5.10 – 5.12 to determine the likelihood of occurrence of the different severities of failure in interacting with the risk assessed agent in post-interaction start time slot 't2'. This process is repeated when the FFL of the risk assessed agent is determined in each time slot of the post-interaction start time phase.

Case 5.7.2: The determined FailureLevel (PFL) of the risk assessed agent in the pre-interaction start time phase has seasonal characteristics, and is the same in all the time slots of that time phase as shown in Figure 5.17.

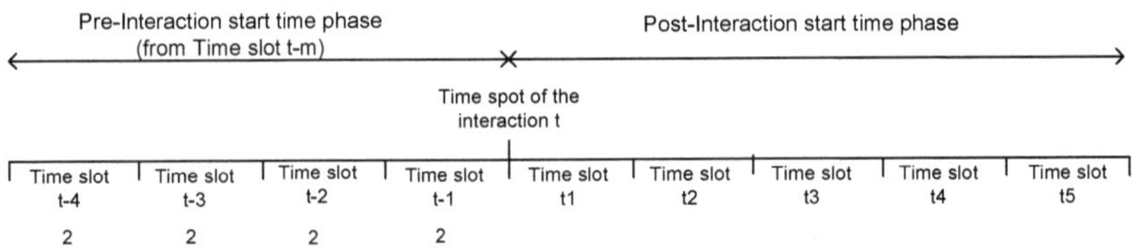

Fig. 5.17 Showing the seasonal trend in the PFL of agent 'B'

In such case, the FailureLevel (FFL) of the risk assessed agent in the time slots of the post-interaction start time phase will be the same as that determined in the pre-interaction start time slots.

Scenario 5.8: The post-interaction start time phase of the risk assessing agent's interaction with the risk assessed agent extends to a point of time in the future, but the time spot is in the current period of time.

If the post-interaction start time phase of the risk assessing agent's interaction is as shown in Figure 5.18, then the risk assessing agent can ascertain the FailureLevel (FFL) of the risk assessed agent in time slot 't1' by using the methodology proposed in scenario 5.6 if he has either a past interaction history with the risk assessed agent in that time slot, or he receives recommendations from other agents in all the assessment criteria of the expectations in that time slot. If the risk assessing agent does not have any of these, then he can utilize the methodology proposed in scenario 5.7 to ascertain the FailureLevel (FFL) of the risk assessed agent in all the time slots of the post-interaction start time phase.

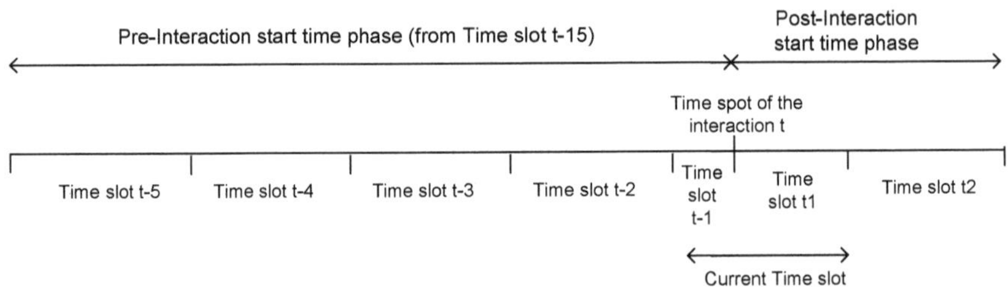

Fig. 5.18 The time slots and the post-interaction start time phase of the interaction

Once the risk assessing agent ascertains the FailureLevel (FFL) of the risk assessed agent in each time slot of the post-interaction start time phase, then he should ascertain the 'FailureLevel Curve' of the interaction in order to quantify the performance risk in interacting with the risk assessed agent.

5.13 Determining the FailureLevel Curve (FLC) of the Business Activity

The 'FailureLevel Curve' (FLC) of the business activity quantifies and represents the performance risk in interacting with the risk assessed agent. This is based on his FailureLevel (FFL) determined during the time period of the post-interaction start time phase. In other words, to the risk assessing agent, the FailureLevel Curve represents the likelihood of the occurrence of different levels of severity of failure during the time period of his interaction with the risk assessed agent, according to the expectations. Hence, the FailureLevel Curve is such that the abscissa of the curve indicates the level or severity of failure from the Failure Scale and the corresponding ordinate or impulse shows the probability or likelihood of occurrence of that level.

A point to be noted here is that the FailureLevel Curve of the interaction is determined by considering only the post-interaction start time phase of the time space. It does not take into consideration the pre-interaction start time phase. This is because the risk assessing agent wants to analyze the transactional risk in interacting with a risk assessed agent during the time in which he possibly interacts with him which is represented by the post-interaction start time phase of the time space.

Two scenarios are possible when ascertaining the FailureLevel Curve of the interaction. They are:

Scenario 5.9: The post-interaction start time phase of the risk assessing agent's interaction with the risk assessed agent is limited to the current time slot in which he is at present.

If the post-interaction start time phase of the interaction is limited to the current time slot as shown in Figure 5.8, and if the risk assessing agent ascertains the FailureLevel (FFL) of the risk assessed agent by utilizing either his past-interaction history or by soliciting recommendations from other agents, or a combination of both as discussed in scenario 5.6, then the determined FailureLevel (FFL) of the risk assessed agent in the post-interaction start time phase is a crisp value on the Failure Scale. In such cases, the FailureLevel Curve (FLC) of the interaction would represent just the determined FailureLevel (FFL) on the abscissa and its corresponding ordinate represents the probability or likelihood of occurrence of that level, which in such cases is 1.

For example, considering the logistics interaction scenario of agent 'A's interaction with agent 'B' is according to the time scenario shown in Figure 5.8 and agent 'A', by utilizing both his past interaction history and recommendations from other agents, ascertains the FFL of agent 'B' as 1 on the Failure Scale, then the FailureLevel Curve of the business association is as shown in Figure 5.19.

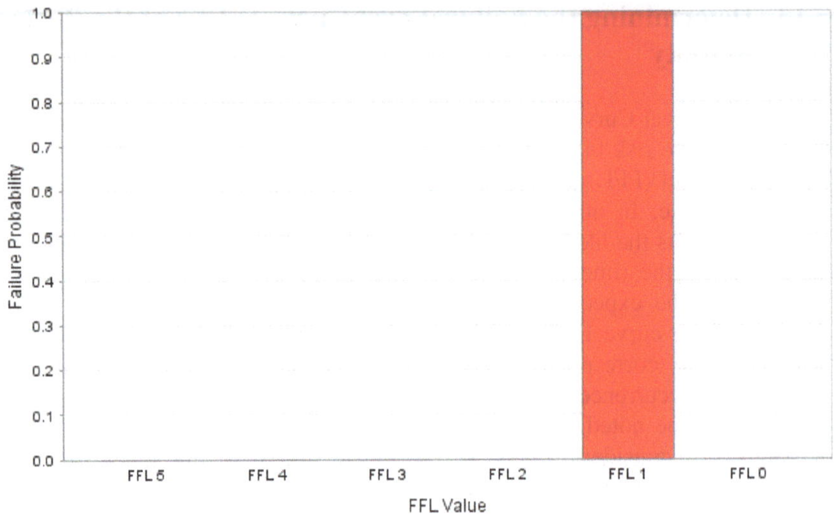

Fig. 5.19 The FailureLevel Curve when the interaction is limited to the current time slot

Scenario 5.10: The post-interaction start time phase of the risk assessing agent's interaction with the risk assessed agent extends to a future point in time.

If the time spot or the post-interaction start time phase extends to a future point in time as shown in Figure 5.17, and if there is a seasonal characteristic in the risk assessing agent's FailureLevel (PFL) in the pre-interaction start time phase as mentioned in case 5.7.2, then the FailureLevel Curve (FLC) of the interaction would be determined as mentioned in scenario 5.9. In contrast, if the FailureLevel (PFL) of the risk assessed agent shows variability (either stochastic variation or trends in variation) as mentioned in case 5.7.1, then the FailureLevel Curve (FLC) of the interaction is plotted by constructing the probability histogram of the sum of the probability of occurrence of each FailureLevel over the time slots of the post-interaction start time phase divided by the number of time slots within that time phase. An example of the FailureLevel Curve of the interaction determined in such a scenario is shown in Figure 5.20.

The risk assessing agent can consider the FailureLevel Curve (FLC) which represents the performance risk in interacting with a risk assessed agent, and utilize it to determine the other subcategory of perceived transactional risk in interacting with it, i.e. the financial risk. In the next chapter, we propose a methodology by which the risk assessing agent can ascertain the financial risk to his resources if interacting with a risk assessed agent.

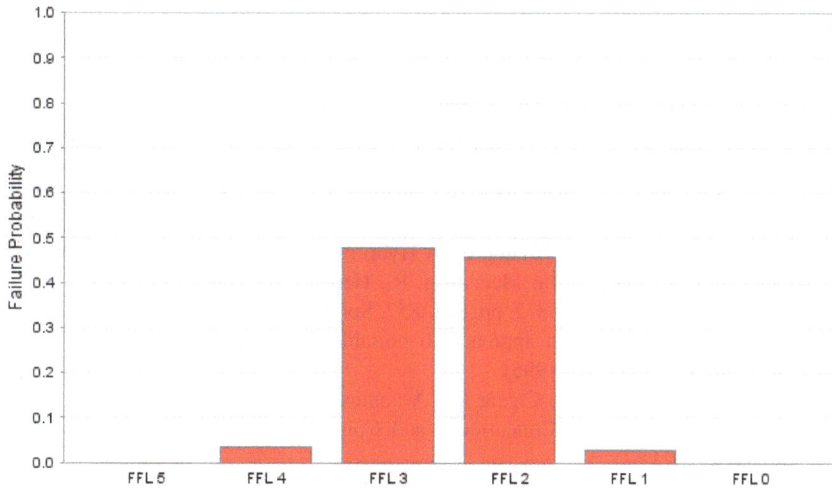

Fig. 5.20 The FLC of the interaction when there are multiple time slots

5.14 Recapitulation

In this chapter, we proposed a methodology by which the risk assessing agent can ascertain a subcategory of perceived transactional risk for its analysis, namely the performance risk. The proposed methodology has four main steps. The first step involves the risk assessing agent identifying the time scenario of the business interaction. The second step involves the risk assessing agent ascertaining the time decay factor, and classifying the recommendations according to their credibility. The third step involves the risk assessing agent ascertaining the FailureLevel of the risk assessed agent in the different time slots and time phases of the time space. In the fourth step, the risk assessing agent ascertains the FailureLevel Curve in his interaction. We defined the Failure Scale which shows the different levels of severity of failure possible in an interaction, and then explained in detail all of the abovementioned steps and the various scenarios accompanying each of them, for the risk assessing agent to ascertain the performance risk in interacting with the risk assessed agent. In the next chapter, we propose a methodology by which the risk assessing agent, by utilizing the performance risk in interacting with the risk assessed agent, can ascertain the other subcategory of perceived transactional risk, namely financial risk in interacting with that agent.

References

1. Hussain, O.K., Chang, E., Hussain, F.K., Dillon, T.S.: A methodology to quantify failure for a risk-based decision support system in digital business ecosystems. Data & Knowledge Engineering 63, 597–621 (2007)

2. Hussain, O.K., Chang, E., Hussain, F.K., Dillon, T.S.: Determining the Failure Level for Risk Analysis in an e-Commerce Interaction. In: Dillon, T.S., Chang, E., Meersman, R., Sycara, K. (eds.) Advances in Web Semantics I. LNCS, vol. 4891, pp. 290–323. Springer, Heidelberg (2008)
3. Chang, E., Dillon, T., Hussain, F.K.: Trust and Reputation for Service-Oriented Environments. John Wiley & Sons, Ltd., West Sussex (2006)
4. Sowa, J.F.: Semantic Networks. In: Shapiro, S.C. (ed.) Encyclopedia of Artificial Intelligence. Wiley (1992)
5. Dong, H., Hussain, F.K., Chang, E.: A Hybrid Concept Similarity Measure Model for Ontology Environment. In: Meersman, R., Herrero, P., Dillon, T. (eds.) OTM 2009 Workshops. LNCS, vol. 5872, pp. 848–857. Springer, Heidelberg (2009)
6. Gruber, T.: A translation approach to portable ontology specifications. Knowledge Acquisition 5, 199–220 (1995)
7. Dong, H., Hussain, F.K., Chang, E.: A context-aware semantic similarity model for ontology environments. Concurrency and Computation: Practice and Experience 23, 505–524 (2011)
8. Hussain, O., Dong, H., Singh, J.: Semantic Similarity Model for Risk Assessment in Forming Cloud Computing SLAs. In: Meersman, R., Dillon, T., Herrero, P. (eds.) OTM 2010. LNCS, vol. 6427, pp. 843–860. Springer, Heidelberg (2010)
9. Rada, R., Mili, H., Bicknell, E., Blettner, M.: Development and application of a metric on Semantic Nets. IEEE Transactions on Systems, Man and Cybernetics 19, 17–30 (1989)
10. Weiss, N.A., Holmes, P.T., Hardy, M.: A Course in Probability. Addison-Wesley (2005)

Chapter 6
Risk Assessment Phase: Financial Risk Assessment in Business Activities

6.1 Introduction

In the previous chapter, we developed the process of perceived transactional risk assessment in business activities. The presented approach for performance risk assessment considers the characteristics and time specific nature of the business activity and models the different levels of failures, along with their probability of occurrence in the successful achievement of the risk assessing agent's expectations. The next step in the process of transactional risk assessment is ascertaining the consequences or the impact of failure of the business association. As mentioned in Chapter 2, this analysis will inform the risk assessing agent how and to what degree he will be affected or impacted as a result of failure to achieve the expectations of the business activity, which will form an important basis for risk assessment.

In this chapter, we present an approach by which the risk assessing agent can determine the consequences or impact (termed financial risk) in the failure of his business association with a risk assessed agent. In addition to assessing performance risk, financial risk too should be ascertained strictly according to the specific characteristics, expectations and time period of the business activity. In the next sections, we present our approach by which these characteristics of transactional risk are captured and the financial risk in a business association determined.

6.2 Identifying the Object of Analysis at Stake to Ascertain the Consequences of Failure

In order to determine the consequences of failure in the business activity, it is important to first identify the *object of analysis* that is at stake in differing business contexts. This can be more fully explained using the following examples of different activities.

Activity 1: A group of aircraft maintenance engineers have been assigned to perform a D-Check on an airplane. The consequences of failure of this activity

O.K. Hussain et al.: Risk Assessment & Management in the Networked Economy, SCI 412, pp. 151–185.
springerlink.com

may be the loss of passenger lives as a result of a crash due to faulty maintenance by the engineers in not completing the tasks expected of them.

Activity 2: Organization 'O' selects and awards a building contract to company 'A' to construct a bridge to a remote island where oil has been discovered. Once complete, Organization 'O' envisages this bridge will play a key role in transporting oil in trucks to the main island which will then be processed to generate revenue and earn profits. The consequences of failure of this activity may be that organization 'O' does not generate the expected revenue from the bridge which it had hoped to generate over a given period of time either due to the bridge being inoperable (that is, not complete) or being closed after completion (that is, it is considered unsafe due to the formation of cracks).

Activity 3: User 'A' is contemplating buying shares in a newly floated public company 'XYZ' in the hope of making a sound investment and obtaining consistent returns. The consequences of failure of this activity may be the financial loss which user 'A' may experience in selling the shares at a lower price as a result of the shares not performing as expected.

Activity 4: User 'B' orders some materials from a manufacturer, that are expected to be delivered just before he travels overseas for a business meeting. Suppose user 'B' does not receive the materials on time as a result of which he has to cancel his trip that may have resulted in him securing a lucrative contract. The consequences of failure to user 'B' are the adverse impacts on his business as a result of not securing the business contract. Also, from the manufacturer's point of view, the consequences of it not delivering the materials on time may result in paying financial compensation to user 'B' and/or the loss of reputation among his customers due to not meeting the delivery deadline.

Activity 5: User 'B' needs urgent medical attention on his finger and has to make a choice between Hospital 'A' and Hospital 'C'. User 'B' knows that Hospital 'A' is under-staffed and he will have a long waiting time before he is attended by the medical staff. The possible consequences or impact to user 'B' in going to Hospital 'A' may be further complications to his finger due to the spread of the infection as a result of delayed treatment.

In the above examples, in which the consequences of failure in each business activity are elaborated, the object of analysis at stake varies according to the context and purpose of each activity. Further, even in an activity, the object of analysis at stake varies according to the different contexts in which the activity is being conducted. For example in Activity 1, one consequence of failure to successfully complete the D-Check on an aircraft is the loss of human life due to a crash. However, from a different viewpoint, another consequence of failure to the company in a different scenario is financial loss as a result of the aircraft being out of service due to it being defective. In activity 2, one consequence of failure to Organization 'O' in not having an operational bridge is financial loss due to suspension of trade, whereas another consequence of failure of the same activity is the loss in Organization 'O's reputation as a result of not being able to meet the oil demands of companies with whom it had business contracts.

So, before determining the consequences of failure of an activity, it is important to first identify the *object of analysis* at stake according to the context in which the activity is being conducted. In the context of the discussion of this book, we define object of analysis as that object, from the risk assessing agent's viewpoint, that is at stake in the business association and for which he fears its loss. Furthermore, it is in relation to this object of analysis that the consequences of failure of a business association have to be determined. However, it is important to note that not all objects of analysis can be quantified in terms of the consequences of failure. In such situations, the consequences of failure should be determined in the subsequent object which the un-quantifiable object of analysis may lead to or in light of the results to which the unquantifiable object of analysis leads. For example, in activity 2, one of the consequence to Organization 'O' in not having the bridge operational may be the loss of reputation among its clients due to not meeting their orders. But how does Organization 'O' determine the impact of the loss to its reputation? It is difficult to determine and quantify the impact of this consequence in its current state, but when a loss in reputation results in a loss in revenue by, for example, lawsuits that may arise due to the failure of activity or a decline in new business contracts, then it can be quantified and determined.

In the next sub-section, we define the object of analysis in which the consequences of failure as a result of failure of the business activity arising from perceived transactional risk analysis is determined.

6.2.1 Defining the Object of Analysis for Financial Risk Assessment in Business Activities

As discussed in Chapter 1, ever developing technologies and the advancement of the web have provided users with various functionalities which facilitate the way business interactions are being carried out in the modern world. But along with the provision of increased functionalities also comes the fear of 'loss' due to not achieving what is desired in an interaction. In the domain of a business activity, we consider that the risk assessing agent wants to achieve certain desired outcomes, and in order to achieve these, it interacts with a risk assessed agent, who has the capability to provide them. In exchange, the risk assessed agent receives an object of mutual consent from the risk assessing agent, as decided between them, in exchange for the services that he promises to provide to fulfil the risk assessing agent's outcomes in the interaction. The object which is exchanged between the two interacting agents can vary and should be pre-determined before initiating the interaction. In the context of the discussion of this book, we consider that a risk assessed agent interacts with a risk assessing agent in order to give it his desired outcomes, which he is capable of providing, as decided in the expectations, in exchange for the pre-determined financial value, as shown in Figure 6.1.

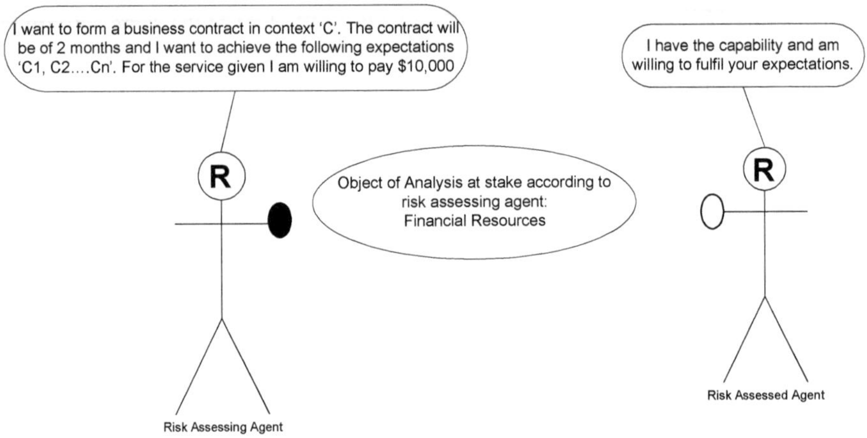

Fig. 6.1 Defining the Object of Analysis at Stake in Business Activities

Hence, the object which the risk assessing agent will have at stake while interacting with a risk assessed agent is its monetary financial value (an agreed financial amount). Subsequently, the loss which the risk assessing agent could experience in the interaction is in the monetary financial value, and the consequences of failure in interacting with a risk assessed agent should be determined in relation to this. The term monetary financial value refers to the resources which the risk assessing agent invests in the business activity to attain his expectations and the term *Financial Risk* of the business activity refers to the consequence of failure to those resources.

To avoid any misunderstandings in the business interaction, we consider that the resources which the risk assessing agent invests in the business association are pre-determined and mutually agreed upon by both agents. This is done when the expectations of the association are formed where the assessment criteria are formalized (Section 4.2.2). Subsequently, when the financial risk in the business activity is being determined, it should be done in relation to the *decided* resources and how they are at *stake* in the business activity.

In the above discussion, two variants of the term *resource* are used, namely resources *decided* and resources *at stake*. Each variant of the term expresses a different meaning that is important to understand and classify for financial risk assessment in a business activity. We discuss the importance of these terms and classify them in the context of the business activity in the next section.

6.3 Differentiating between Resources Decided in Expectations and Resources at Stake

As discussed in the previous chapters, the perceived transactional risk which the risk assessing agent may have to analyze in forming an association with the risk assessed agent could be in a set of activities over a given finite period of time, or in a single activity which extends over a finite period of time. Furthermore, the time period of the risk assessing agent's activity with the risk assessed agent might be limited to the current period of time (Figure 5.8) or it might extend to a point of time in the future (Figure 5.7). To investigate each scenario while undertaking financial risk assessment, the risk assessing agent should follow the division of time as it does when determining the performance risk in forming an association with a risk assessed agent. However, an important point to note here is that the business association of the risk assessing agent with the risk assessed agent occurs only in the post-interaction start time phase of the time space (Figure 5.7). In other words, this is the time phase where the risk assessing agent will invest the *decided* resources and have these *at stake.* So, while undertaking financial risk assessment, he needs to consider only that phase of the time space and not the combination of the pre-interaction and post-interaction start time phases as is the case when undertaking performance risk assessment. This is because while forming the expectations, the risk assessing agent *decides* on the amount of resources he will invest and how he will have those resources *at stake*. This differs to performance risk assessment where the risk assessing agent has to consider the past interaction history of the risk assessed agent before predicting his performance risk in the future, and therefore needs to consider both time phases.

To differentiate and explain the importance of the words *decided* to invest and *at stake* in the previous discussion, let us consider the business association example between risk assessing agent 'A' and logistics company (risk assessed agent 'B') from Section 4.2.2.1. Let us assume that there are 5 time slots in the post-interaction start time phase ('t1'-'t5'), and risk assessing agent 'A' decides with the risk assessed agent in the expectations to invest 10,000 euros in the interaction to achieve his desired outcomes, according to the assessment criteria of the expectations. There are many ways by which the risk assessing agent can invest the determined financial resources, as discussed in scenarios 6.1 and 6.2.

Scenario 6.1: Variable Investments by the risk assessing agent in each time slot

The risk assessing agent invests its resources variably in the post-interaction start time phase of its interaction as shown in Figure 6.2. The resources he invests in each time slot are 500, 2,000, 2,500, 2,000 and 3,000 euros in time slots 't1', 't2', 't3', 't4' and 't5' of the post-interaction start time phase, respectively, as shown in Figure 6.3.

Scenario 6.2: Equal investments by the risk assessing agent in each time slot

The risk assessing agent invests his resources equally in the post-interaction start time phase of his interaction, as shown in Figure 6.4. The resources invested are 2,000, 2,000, 2,000, 2,000 and 2,000 euros in time slots 't1', 't2', 't3', 't4' and 't5' of the post-interaction time phase respectively, as shown in Figure 6.5.

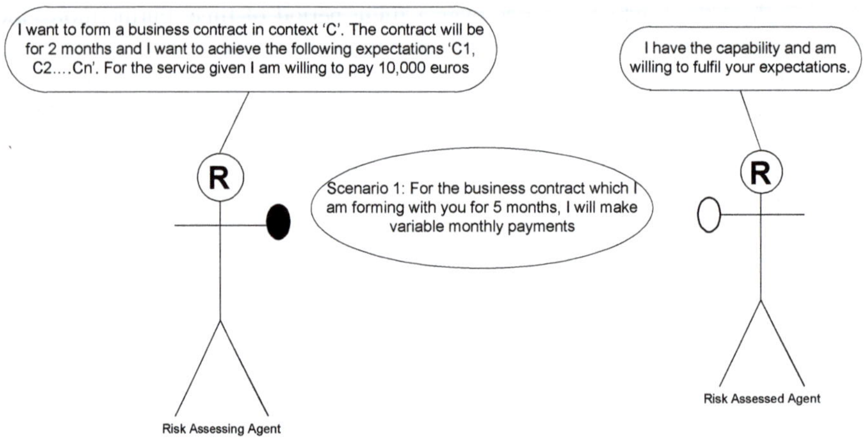

Fig. 6.2 Variable investment of resources in the business interaction

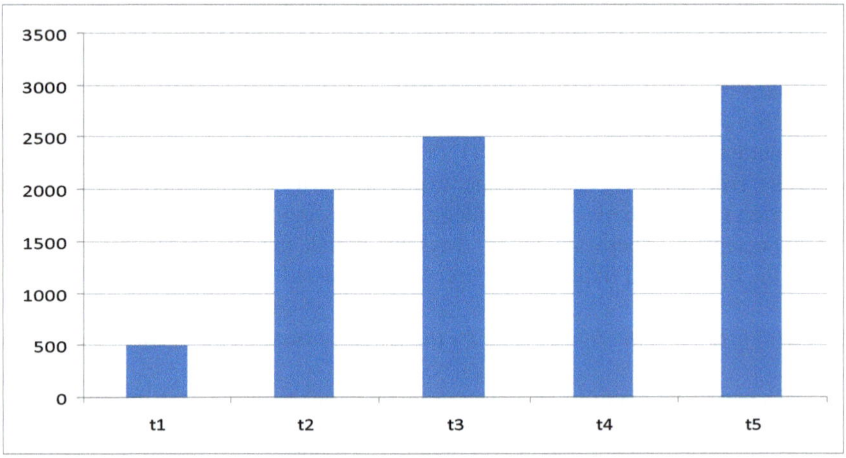

Fig. 6.3 Investment in each time slot according to the type of investment described in scenario 6.1

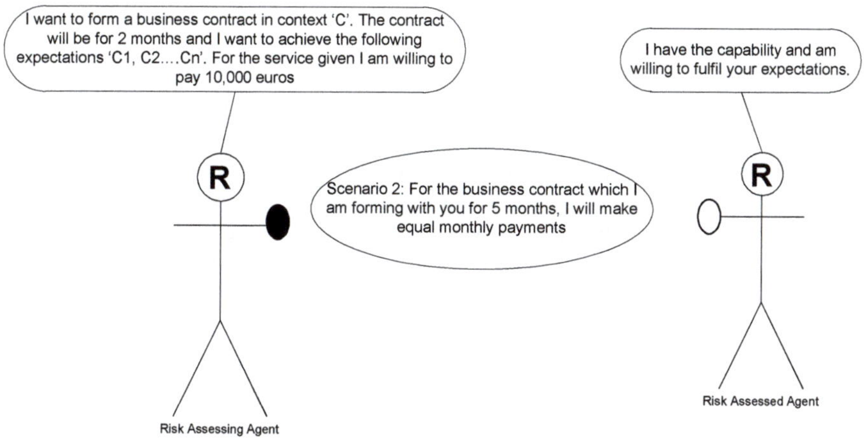

Fig. 6.4 Equal investment of resources in the business interaction

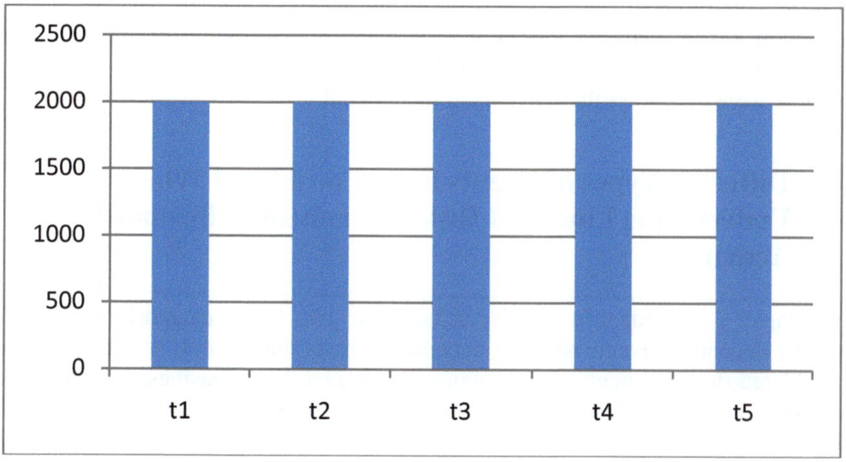

Fig. 6.5 Investment in each time slot according to the type of investment as described in scenario 6.2

In both the abovementioned scenarios, even though the total financial resources invested in the business association is 10,000 euros, at no time during time slots 't1' – 't4' are the net financial resources of 10,000 euros belonging to the risk assessing agent invested and at *stake*, as decided in the expectations. In other words, even though the total financial resources *decided* in the expectations are 10,000 euros, the amount *at stake* at any given point of time in each scenario varies according to the type of investment being made by the risk assessing agent, only reaching the total financial amount decided in the expectations in the last time slot of the business activity.

So, there is a difference between the resources *decided* in the expectations and the net resources *at stake* at any given point of time in the interaction. The resources decided in the expectations represents the total net financial amount which the risk assessing agent will invest to achieve his outcomes in interacting with the risk assessed agent, whereas the 'resources at stake' represents the value of the amount at stake from the net resources that the risk assessing agent invests throughout the duration of his interaction with the risk assessed agent, according to the type of investment it makes. With the variation in the investment nature of the financial resources in the interaction, the amount of the risk assessing agent's resources *at stake* at any given point of time in the interaction varies, as does the risk of financial loss to the risk assessing agent. Therefore, it is important to capture this variability in the financial resources *at stake* in addition to considering the financial resources *decided* in the expectations to assess the overall financial risk in the interaction.

Once the resources at stake in a business interaction have been determined, the next step in assessing financial risk is to identify the different types of events and their associated uncertainties that will lead to the risk assessing agent not achieving its desired outcomes from the resources at stake in the business activity. It is important to note that the identified events are specific to the business activity and have to be analysed carefully according to the context of discussion or analysis of the business activity and the objectives to be achieved. In the next section, we will discuss these events in more detail.

6.4 Different Types of Events to Be Considered When Undertaking Financial Risk Assessment in Business Activities

In Chapter 2, we identified and discussed the two broad categories of events, namely dependent events and non-dependent events that need to be identified and considered during perceived transactional risk analysis in business activities. As discussed in Chapter 5, during performance risk analysis in a business activity, the risk assessing agent considers and analyses only those events from his expectations for which he has to depend on the risk assessed agent. In other words, he only considers the dependent events and ascertains the probability of the risk assessed agent not committing to them according, to his expectations. The other type of events, namely non-dependent events, are outside the scope of dependence of the risk assessed agent, but may have an impact on the successful completion of the business activity if they are not achieved according to the expectations. The following two scenarios illustrate the importance of considering such events during perceived transactional risk analysis:

Scenario 6.3: Let us consider the interaction scenario discussed in Section 2.6 where Agent 'A' interacts with a logistics company to transfer his goods from Europe to Australia. According to the expectations, Agent 'A' has to make a payment of 1,500 euros when the goods are picked up and another part payment of

1,500 euros in Australian dollars when the goods are delivered to the destination address in Perth. It is possible that after forming the expectations, the exchange rate of the Australian dollar against the Euro varies which may result in agent ‘A’ paying more financial resources to the logistics company than what was initially decided in the expectations. The variation in the occurrence of this event is not dependent on the logistics company but on other factors, hence is a non-dependent event. Agent ‘A’ can consider such extra investment which he now has to make as opposed to what was expected and initially decided as his financial loss in the business activity. Subsequently, while determining the financial risk, agent ‘A’ should also consider the uncertainty of such types of events too.

Scenario 6.4: Cloud computing has enabled users to access various resources and applications from third party users without actually having to purchase them. The users consume resources as a service and in return, pay the provider only for the time they use them. Business contracts are formed between the users and service providers to ensure that the required services and applications that are of great importance to the users are available to them as and when needed. Let us consider the scenario of agent ‘A’ forming a business contract with service provider ‘B’ to access resource ‘D’ for an hour each day on provider ‘B’s virtual cloud environment for a price of A$500 per day. Let us consider that to access resource ‘D’, agent ‘A’ has to utilize a higher bandwidth of network and in order to do this, he takes an internet plan that allows him to access an hour of high network bandwidth usage for free and then pay for any usage above that time limit. Let us consider that on a particular day, due to congestion on servers, agent ‘A’ has to utilize the network above his daily time limit, incurring costs in excess of his monthly plan. These extra financial resources that agent ‘A’ might have to pay can be considered as ‘loss’. Subsequently, apart from simply considering the uncertainty of the dependent events (from service provider ‘B’), agent ‘A’ should also consider the uncertainty of the non-dependent or consequent events in this case in order to be able to make an accurate assessment of financial risk in forming a business association.

The above discussion provides a brief insight on non-dependent uncertain events and the impact or consequences that they can have on the successful completion of the business activity, in addition to dependent events. It can be argued that the occurrence of non-dependent events will not always lead to consequences that are negative, but may also lead to situations that are profitable or positive to the risk assessing agent. For example, it is possible that the Australian dollar may have a high exchange rate against the European euro in scenario 6.3; or agent ‘A’ may complete his daily tasks in 45 minutes and save the unused network bandwidth for later use in the month in scenario 6.4. In our approach, we consider the occurrence of all such possible scenarios, but during the analysis of financial risk, we only consider the occurrence and impact of events that will lead to a negative outcome, as our focus and aim in this chapter is to determine the possible financial losses in a business activity as an impact of risk.

In the next section, we list the series of steps to be followed for financial risk assessment in business activities.

6.5 Steps to Determine Financial Risk in Business Activities

According to the formed expectations of the business activity, the risk assessing agent assumes to attain his objectives in interacting with a risk assessed agent by the investment of his resources as decided in the expectations. However due to the uncertainty associated with both dependent and non-dependent events, this may not happen, resulting in financial risk in the business activity. In order to determine the financial risk in the business activity, we propose the following steps to be followed:

1. *Accurately identify the amount of resources at stake in the business activity.* In our approach, this is done by plotting the *Amount Invested Curve*. The process of determining the Amount Invested Curve is discussed further in Section 6.7.
2. *Determine the uncertainty that will lead to variations in the occurrence of dependent and non-dependent events from the desired level.* The variation of dependent events in the business activity from that decided in the expectations is represented by performance risk (as discussed in Chapter 5). The variation of non-dependent events in the business activity from that decided in the expectations is captured and determined by plotting the *Variation Curve*. The process of determining the Variation Curve is discussed further in Section 6.9.
3. *Determine the probability of the risk assessing agent not achieving the desired outcomes from the resources decided in the expectations.* As a result of the uncertainty that will lead to variations in the occurrence of dependent and non-dependent events from the desired level (step 2), it is possible that the risk assessing agent may not achieve the desired outcomes *from the resources* decided in the expectations (step 1). Furthermore, this also implies that the risk assessing agent may have to keep extra financial resources at stake. This is determined by plotting the *Factual Amount Invested Curve, Resources to be kept at Stake Curve* and *Total Factual Amount Invested Curve* of the business activity. These curves ascertain the impact of the performance risk (dependent events) and variation curve (non-dependent events) on the resources at stake in the business activity. The process of determining these curves is discussed further in Sections 6.8, 6.10 and 6.11, respectively.
4. *Utilize the analysis determined in step 3 to ascertain the financial risk to the risk assessing agent in the business activity.* Using the analysis in step 3 which represents the extra financial resources needed to be kept at stake along with the probability of those extra resources needing to be kept at stake, plot the *Loss Curve* of the business activity which represents the financial risk to the risk assessing agent in the business activity. The process of determining the Loss Curve is discussed further in Section 6.12.

In the next sections, we present each of the above mentioned steps in detail by which the risk assessing agent can determine the financial risk in the business activity.

6.6 Definition of Key Terms Used While Undertaking Financial Risk Assessment

In this section, we define the key concepts that will be used while undertaking financial risk assessment in business activities.

6.6.1 Amount Invested Curve (AIC)

Amount Invested Curve (AIC) accurately represents the amount of the risk assessing agent's resources at stake from the net resources decided in the expectations.

6.6.2 Factual Amount Invested Curve (FAIC)

Factual Amount Invested Curve (FAIC) is an extension of the AIC and represents the level of resources to be kept at stake by the risk assessing agent, taking into consideration the performance risk of the risk assessed agent.

6.6.3 Variation Curve (VC)

Variation Curve (VC) represents the different levels of variation in the occurrence of a non-dependent event from what was decided in the expectations.

6.6.4 Resources Decided at Stake Curve (RDC)

Resources Decided at Stake Curve (RDC) accurately represents the amount of the risk assessing agent's resources at stake from the net resources decided in the expectations on which variations in the occurrence of a non-dependent event will have an impact.

6.6.5 Resources to Be Kept at Stake Curve (RKC)

Resources to be Kept at Stake Curve (RKC) is an extension of the RDC. The RKC represents the amount of resources to be kept at stake by the risk assessing agent, taking into consideration the different positive level/s of variation in the occurrence of non-dependent event/s from what was decided in the expectations.

6.6.6 *Extra Investment Curve (EIC)*

Extra Investment Curve (EIC) represents the extra level of financial resources to be kept at stake by the risk assessing agent by considering the different positive level/s of variation in the occurrence of non-dependent event/s from what was decided in the expectations.

6.6.7 *Total Factual Amount Invested Curve (TFAIC)*

Total Factual Amount Invested Curve (TFAIC) is an extension of the FAIC. It combines and represents the amount of financial resources which the risk assessing agent has to keep at stake by considering the performance risk of the risk assessed agent (FAIC) and the extra level of financial resources to be kept at stake by considering the different positive level/s of variation in the occurrence of non-dependent event/s from what was decided in the expectations (EIC).

6.6.8 *Loss Curve (LC)*

Loss Curve (LC) represents the cumulative probability of the extra financial resources which the risk assessing agent has to keep at stake by considering the variation in the occurrence of both dependent and non-dependent events from what was decided in the expectations.

6.7 Ascertaining the Amount Invested Curve (AIC) of the Business Activity

As discussed earlier, the variability in the nature of the risk assessing agent's investment of resources may have elements of stochastic variation, trends in variation or seasonality in variation; each of which will lead to variation in the financial resources at stake. To capture this variation, we propose that the risk assessing agent should:

(a) determine the investment pattern of its financial resources in the post-interaction start time phase of the interaction, as shown in Figure 6.3, and
(b) accurately determine the value of a financial amount at stake throughout the interaction time period from the net invested resources.

This is best calculated by using probability theory and plotting the AIC. The AIC of a business activity is ascertained by considering the investment nature of the risk assessing agent and by using the cumulative probability mass function to determine the probability of the net worth of the business activity to be at least, equal to, or greater than a certain amount through the post-interaction start time slots. Mathematically, the AIC of a business activity can be determined by using the following function [1]:

$$P(x) = \frac{m}{n} \tag{6.1}$$

where:

x= the financial amount on which the AIC is being determined,
m= number of time slots in which the financial amount 'x' will be invested and at stake,
n= total number of time slots in the time space.

Alternatively, equation 6.1 can also be used to determine the cumulative probability of each amount at stake throughout the time period of the business activity. So, depending on the way the AIC of a business interaction is plotted, it either represents:

(a) the probability of a financial amount invested and at stake from the net resources in each time slot of the post-interaction start time phase, or
(b) the probability of the cumulative financial resources at stake at the end of each time slot of the post-interaction start time phase.

The AIC gives the risk assessing agent an accurate determination of his resources at stake as opposed to the net resources that were decided in the expectations [2]. To illustrate, let us consider the investment pattern discussed in scenarios 6.1 and 6.2. The AIC, according to these investment patterns, is shown in Figures 6.6 and 6.7, respectively.

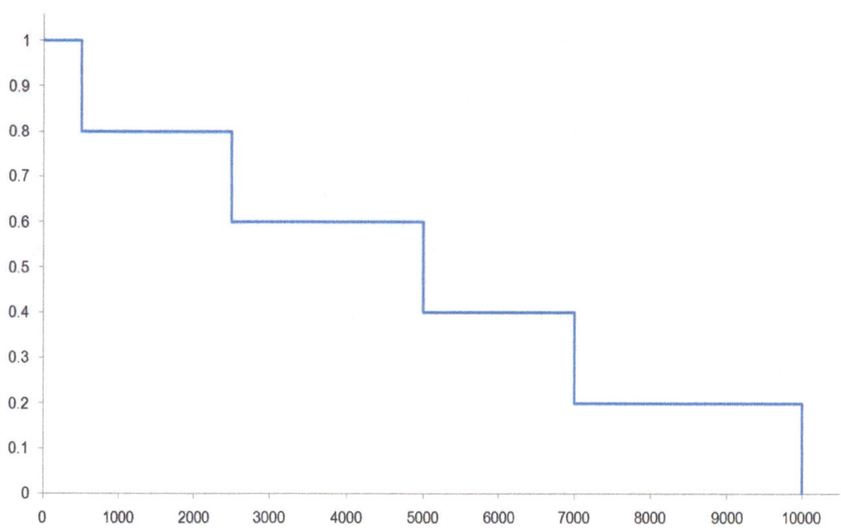

Fig. 6.6 The AIC according to the resources invested in scenario 6.1

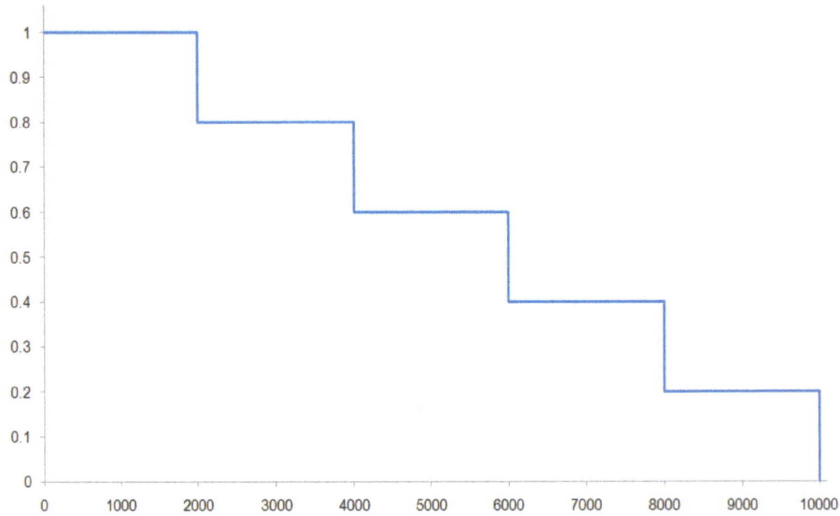

Fig. 6.7 The AIC according to the resources invested in scenario 6.2

In both figures 6.6 and 6.7, the x-axis represents the net financial resources and the y-axis represents the probability of those resources at stake in the business activity. As can be seen from Figures 6.6 and 6.7, even though the net financial resources of the risk assessing agent at stake in both investment patterns is 10,000 euros, the probability of the net value of resources being at stake varies according to the type of investment in the business activity. Such variation is captured by plotting the AIC of the business activity.

According to the expectations of the business activity, the risk assessing agent assumes to attain his objectives by the investment of his resources, as represented in the AIC. However, due to uncertainty associated with variations due to the occurrence of dependent and non-dependent events, this may not happen, resulting in the risk assessing agent keeping extra resources at stake, apart from those decided in the expectations. These extra resources and the probability of the risk assessing agent having to keep them at stake need to be determined to ascertain the financial risk in the business activity. In the next section, we propose an approach by which the extra resources to be kept at stake due to uncertainty in the commitment of dependent events as decided in the expectations can be determined.

6.8 Ascertaining the Factual Amount Invested Curve (FAIC) of the Business Activity

To determine the amount of extra resources to be kept at stake due to uncertainty associated with the occurrence of dependent events, the impact of the performance risk of the risk assessed agent (FLC) on the resources initially decided to be at stake in the interaction (AIC) should be determined. As both the AIC and FLC are probabilistic curves, we will utilize probability theory to determine the impact of the FLC on the AIC, and the resulting curve too will be a probabilistic curve. We term the resulting curve of the impact of the AIC and FLC as the *Factual Amount Invested Curve (FAIC).* The FAIC shows the *required resources* and the *required* probability of an amount to be kept at stake from these resources by the risk assessing agent, throughout the duration of the post-interaction start time phase, taking into consideration the risk assessed agent's performance risk. This is in contrast to the AIC which shows the *actual* probability of an amount kept at stake as *decided* in the expectations [2]. Hence, the FAIC is an extension of the AIC and the modification between the two curves reflects the impact of the performance risk of the risk assessed agent on the resources initially decided upon and invested according to the expectations.

There are many techniques which can be utilized to determine the impact of a probabilistic function on the other probabilistic functions. In our approach, we utilize the convolution operator to determine the impact of the FLC of the business association on the AIC. We will discuss the convolution operator and how it is used in our approach in the next sections.

6.8.1 Convolution Operator to Determine the FAIC of the Business Activity

Convolution is a mathematical integral operator which expresses the amount of overlap and impact of one function as it shifts over the other. It has been studied and utilized in various complex scenarios in domains such as power generation [3-5] and signal processing etc, but no work has been done in applying it to the area of risk management in business activities. There are various methods for calculating the convolution of two probabilistic functions. In our approach, we utilize the following two:

1. conventional method, and
2. cumulants method

6.8.2 Conventional Method for Convolution

The conventional method for convolution is a recursive step where one probabilistic function is broken down into different points and then the density function of the other probabilistic function is convolved recursively at each point. In our approach, as both the inputs AIC and FLC are probabilistic curves, we divide the AIC into different points and convolve it with the probability density function of the FLC. Mathematically, the conventional convolution process can be described by the random variables of both the FLC and AIC by:

$$\text{FAIC} = \text{FLC} \oplus \text{AIC} \tag{6.2}$$

where:

FAIC = Factual Amount Invested Curve,
FLC = FailureLevel Curve of the risk assessed agent,
AIC = Amount Invested Curve.

The symbol $\oplus$ represents the convolution operator between two probabilistic functions. Two scenarios arise while determining the FAIC by using equation 6.1.

Scenario 6.5: There is only one severity of failure in the FailureLevel Curve of the risk assessed agent:

If there is only one severity of failure in the FLC of the business activity as shown in Figure 5.19, then the convolution of the FLC with the AIC to obtain the FAIC at each point 'x' is obtained by:

$$\text{FAIC}(x) = p_i * \text{AIC}(x - FL_i) \qquad \text{for } (x - FL_i) \geq 0$$

or

$$\text{FAIC}(x) = p_i \qquad \text{for } (x - FL_i) < 0 \tag{6.3}$$

where:

x = the point at which the FAIC is ascertained,
FL_i = severity of the FailureLevel_i,
p_i = magnitude of occurrence of FailureLevel_i (FL_i),
AIC $(x - FL_i)$ = Amount Invested Curve value at point $(x - FL_i)$.

Scenario 6.6: There is more than one severity of failure in the FailureLevel Curve of the risk assessed agent:

If there is more than one severity of failure in the FLC while interacting with a risk assessed agent as shown in Figure 5.20, then the FAIC as a result of convolution of the FLC and AIC is determined by using the recursive formulae given in equation 6.4

$$\text{FAIC}(x) = \sum_{i=1}^{n} p_i * \text{AIC}(x - FL_i) \qquad \text{for } (x - FL_i) \geq 0$$

or,

$$\text{FAIC}(x) = \sum_{i=1}^{n} p_i \qquad \text{for } (x - FL_i) < 0 \tag{6.4}$$

where:

n = the number of Failure Levels in the FLC,
x = the point at which the FAIC is to be ascertained,

FL_i = magnitude of failure of $FailureLevel_i$,
p_i = magnitude of occurrence of $FailureLevel_i$ (FL_i),
AIC ($x - FL_i$) = Amount Invested Curve value at point ($x - FL_i$).

The conventional method for convolution is best suited to determine the FAIC of the business activity when there is only one level of failure in the FLC. In scenarios where the perceived transactional risk is being determined for a set of activities or in each time slot of the business activity which have different levels of failure in it, then this process becomes computationally expensive. In these situations, the cumulant method of convolution is best suited to ascertain the FAIC by combining the AIC with the FLC. This process is less computationally expensive compared to the conventional method that uses a recursive formula. We discuss the process of convolution by using cumulants in the next section.

6.8.3 Cumulant Method for Convolution

The cumulant method as opposed to the conventional method is determined by the properties known as the moments and cumulants of the probabilistic distribution functions. Moments of a distribution function are termed the expectations of different powers of the random variable. Alternatively, they are a set of descriptive constants which are useful for measuring the properties and in certain circumstances for specifying the distribution function [6]. But they are not the best set or the only set of constants to represent the distribution function. Another series of constants used to represent the distribution function are known as cumulants (represented by κ). They have properties which can better describe the underlying distribution function and which are more useful from a theoretical standpoint. The cumulants of a distribution function are determined by its statistical moments. In the cumulant method, the convolution of the distribution functions is expressed as a sum of the individual cumulants of the random variables representing the distribution functions. Based on the knowledge of the resulting obtained cumulants, the distribution function of the output function is approximated by using the series expansion functions. So, the series of steps to be carried out to ascertain the FAIC by using the cumulant method of convolution are:

1. The probability density function of each input function (FLC and AIC) is determined.
2. The moments and cumulants of the input functions to order 4 are determined.
3. The cumulants of AIC and FLC are added to find the resulting cumulants of the FAIC.
4. A series expansion function is utilized to ascertain the FAIC at that point in time from the knowledge of its cumulants.

There are different available series expansion functions. We give a brief description of the following two functions which we will utilize in our approach. These are:

1. Gram-Charlier Series Expansion
2. Laguerre Series Expansion.

A brief description of the above mentioned three series and how they are utilized to represent the FAIC of the business activity is discussed in the next sub-sections.

6.8.3.1 Gram-Charlier Series Expansion

Gram-Charlier series expansion is a probabilistic series expansion function from the knowledge of its cumulants. The Gram-Charlier series has the normal distribution as its leading term, followed by terms containing successive derivatives of the normal from order 3 onwards. The motivation for this series comes from the central limit theorem. This theorem basically states that a sum of any 'n' independent random variables (satisfying certain conditions) is approximately normal distributed for a sufficiently large n [7]. In Gram-Charlier expansion, certain coefficients are added to the normal to improve the approximation given by the normal distribution. The importance of these terms in the series decreases as the number of random variables involved in the sum increases. The standardized third cumulant of the series is proportional to the skewness of the resulting density function series with respect to the standard normal, whereas the standardized fourth cumulant is proportional to the kurtosis of the resulting density function series. The skewness of the density function shows the direction of its inclination, whereas the kurtosis shows the breadth of the resulting density function. The Gram-Charlier series expansion of the resultant distribution function is represented by:

$$\text{FAIC}(z) = 1 - \int_{-\infty}^{z} N(z)dz + \sum_{k=3}^{5} \frac{G_k}{k!} N^{(k-1)}(z) \tag{6.5}$$

where:

z = standardized variable,
G_k = standard cumulant of order 'k',
N (z) dz = standard normal,
$N^{(k-1)}(z) = (k-1)^{th}$ derivative of the standard normal,

The standard cumulant of order 'k' which is represented by G_k are the cumulants of the FAIC of the interaction obtained as the sum of the resulting cumulants of AIC with the FLC. The standardized variable is given by the expression:

$$z = \frac{x-\mu}{\sigma} \tag{6.6}$$

where:

x = investment level for which the corresponding value on the FAIC has to be determined,
μ = mean of the resulting random variable,
σ = standard deviation of the resulting random variable.

It is sufficient to take only two terms in the Gram-Charlier series expansion for the normal distribution to approximate the FAIC, because as mentioned earlier, the importance of the terms in the series decreases as the number of random variables involved in the sum increases.

6.8.3.2 Laguerre Series Expansion Function

The Laguerre function is a series expansion function from the knowledge of the moments and cumulants of the underlying function. It uses a form of Gamma distribution as the leading term followed by terms containing successive derivatives of this distribution. The derivatives are given by multiplication of the leading term by suitable Laguerre polynomials. The formal definition of this series is given by the probability density function represented by Equation 6.7

$$g(x) = \sum_{k=0}^{\infty} d_k L_k^{(r)}(x)\, h(x) \tag{6.7}$$

where:

d_k represents the coefficient of the expansion
$L_k^{(r)}$ (x) represents the Laguerre polynomial of order 'k' which depends on the variable r.

h (x) is the leading term that is represented by $h(x) = \dfrac{xre^{-x}}{\Gamma(r+1)}$

The coefficients of the expansion depend on the moments as given in equation:

$$j = \frac{(-1)^j}{j!(1+r)_j} \sum_{k=0}^{j} \binom{j}{k} (-1)^k \frac{(1+r)_j}{(1+r)_k} m_k \tag{6.8}$$

where:

m_k= Moment of order 'k' with respect to 0,

$\binom{j}{k} = \dfrac{j!}{(j-k)!k!}$

The notation $(1+r)_j$ can be expanded as (1+r) (2+r) (3+r)...........(j+r)

The Laguerre polynomials can be calculated by using the recursive expression

$$L_{n+1}^{r} = [x-r-1-2n]\, L_n^{(r)}(x) - n(n+r)\, L_{n-1}^{(r)}(x) \tag{6.9}$$

Once the different coefficients have been determined, then the FAIC is the complement of the corresponding distribution function which is:

$$\text{FAIC}(z) = 1\text{-}G(x) = e^{-x} + x\,e^{-x} \sum_{k=1}^{\infty} d_k L_{k-1}^{(1)}(x) \tag{6.10}$$

6.8.4 Determining the Financial Resources to Be Kept at Stake due to the Non-occurrence of the Dependent Events according to the Expectations

By utilizing one of the abovementioned approaches for convolution, the impact of the FLC on the AIC can be determined to plot the FAIC of the business activity. To illustrate with an example, let us consider the interaction scenario of risk assessing agent 'A' forming a business association with the logistics company (agent 'B'), as discussed in Section 6.3. Let us consider that the AIC which shows the resources decided to be kept at stake is as shown in Figure 6.6 and the FLC of agent 'B' to commit to the dependent events of the expectations is as shown in Figure 5.20. Using the conventional method for convolution discussed in the earlier section, the FAIC of the business association interaction is shown in Figure 6.8 [2]. The x-axis of Figure 6.8 represents the net worth of financial resources to be kept at stake due to the non-achievement of dependent events as decided in the expectations and y-axis represents the probability of those resources to be at stake in the business activity.

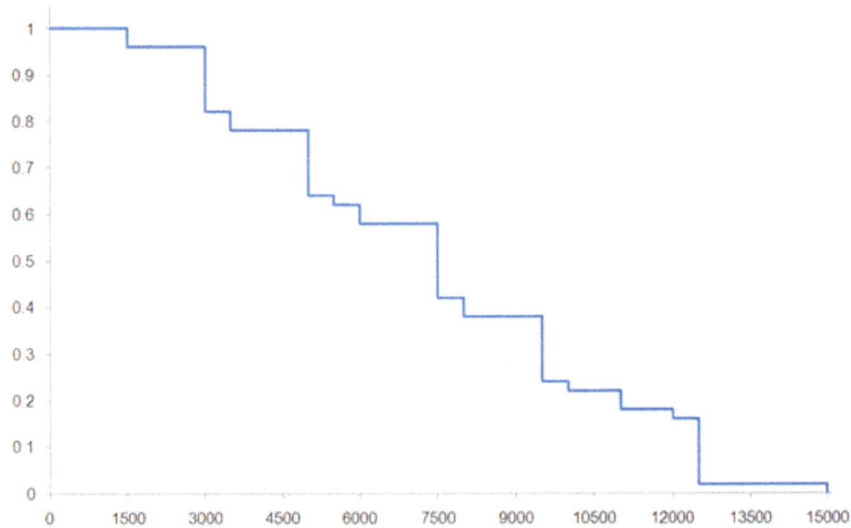

Fig. 6.8 FAIC in agent 'A's business interaction with agent 'B'

As can be seen from Figures 6.6 and 6.8, the FAIC is inflated compared to the AIC due to the additional resources and the probability of those resources needing to be kept at stake by the risk assessing agent, by considering the performance risk of the risk assessed agent. In certain cases, when the risk assessed agent does not have any performance risk in committing to the expectations of the risk assessing agent, then the FAIC of the business activity will be the same as the AIC. This implies that the risk assessed agent will commit to the objectives and assessment criteria of the risk assessing agent within the time period and with the resources as decided in the expectations, thereby nullifying any financial risk. But, in cases where the risk assessed agent might have varying severities of performance risk in committing to the expectations of the risk assessing agent, then the FAIC of the interaction will be inflated, compared with the AIC. The degree of inflation on the FAIC compared with the AIC is dependent on the severity of performance risk of the risk assessed agent. The higher the severity of performance risk of the risk assessed agent, the greater the inflated FAIC as compared with the AIC and vice versa.

Once the risk assessing agent determines the resources to be kept at stake due to uncertainty in the occurrence of dependent events, it should then consider the non-dependent events and conduct a similar analysis to determine the financial resources that it has to keep at stake due to the uncertainty of their occurrence, as decided in the expectations. We discuss our approach to calculate this in the next section.

6.9 Ascertaining the Variation Curve (VC) of the Business Activity

Before the impact of the non-dependent events on a business activity can be determined, variations in their occurrence from that decided in the expectations need to be determined. It is important to note that the non-dependent events are specific to the business activity and hence, have to be identified according to the specific characteristics and expectations of the business activity in which they have to be determined. Further, in our approach, we denote the non-dependent events (NDE) in the business interaction as:

$$\{NDE_1, NDE_2, NDE_3, \ldots\ldots\ldots, NDE_n\}$$

where:
'n' represents the number of non-dependent events identified by the risk assessing agent in a business interaction.

Unlike dependent events, the occurrence of non-dependent events in a business activity may be spontaneous and volatile with no certain pattern of occurrence, and as a result, it is extremely difficult to have a distribution that models trends in their occurrence during the business activity time period. However, on the other hand, as discussed in Section 6.4, it is important for the risk assessing agent to account for the uncertainty of such events in order to consider the expected financial loss which may be incurred as a result while determining the overall financial risk. So, to capture such events, in our approach, we utilize an uncertainty propagation technique that models the different variations in the occurrence of a non-dependent event/s from the desired level in the expectations. One such technique is the Monte Carlo method which is best utilized when the underlying model is complex, nonlinear and involves uncertain parameters which makes it best suited to our problem. It is a sampling method for analyzing uncertainty propagation that uses simulations of independent random numbers and models the uncertainty associated with the defined problem. It does not give a deterministic output at each simulation but iteratively evaluates a deterministic model (usually the problem framework) using sets of random variables within the defined range as inputs [8]. By utilizing the Monte Carlo technique, we propose that the risk assessing agent determines the variations in the occurrence of a non-dependent event. Such analysis should be then be utilized to plot a probabilistic curve, termed the VC.

There are two possible scenarios when the risk assessing agent has to determine the variation in the occurrence of non-dependent events in the business activity. They are:

Scenario 6.7: There is only one non-dependent event identified over the time period of the business association.

Case 6.7.1: The non-dependent event occurs in only one time slot of the post-interaction start time phase.

If there is only one non-dependent event identified and if that event occurs in only one time slot of the post-interaction start time phase, then by utilizing the Monte Carlo technique, the risk assessing agent should be able to determine the different variations in the occurrence of that event in the given time period. This analysis should then be transformed over a range of -100% to 100%, where each value in that range represents:

(a) the degree of variation in the occurrence of a variable from what was decided in the expectations, and
(b) the probability of the occurrence of the level of variation.

A value of 0 for the level of variation of an event represents the occurrence of an event as decided in the expectations, whereas any value above or below that represents a variation in the occurrence of that event. The level of variation in that event from the one decided in the expectations is represented by the numerical value in the range. A representation of the VC is shown in Figure 6.9.

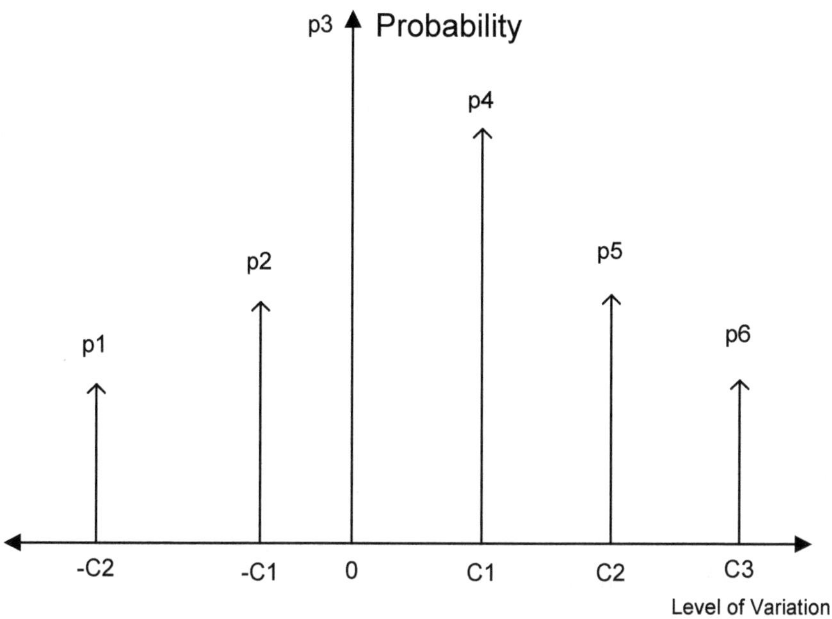

Fig. 6.9 The VC of a non-dependent event over the time period of the business activity

Case 6.7.2: The non-dependent event occurs in more than one time slot of the post-interaction start time phase.

If the non-dependent event occurs in more than one time slot of the post-interaction start time phase, then the risk assessing agent should repeat the steps elaborated in case 6.7.1 and determine the probabilistic curve which represents the different variations of that event along with their probability of occurrence in each time slot. The VC of that event in the post interaction start time phase is determined by constructing a probability histogram of the sum of the probability of occurrence of each level over the time slots divided by the number of time slots in which the non-dependent event occurs.

Scenario 6.8: There is more than one non-dependent event identified over the time period of the business association.

If there is more than one non-dependent event over the time period of the business association, then according to its pattern of occurrence (whether in a single time slot or multiple time slots in the post interaction start time phase) the risk assessing agent should carry out the appropriate steps elaborated in Cases 6.7.1 and 6.7.2 of Scenario 6.7 and plot the VC for each of them as shown in Figure 6.9.

Once the VC for each non-dependent event has been determined, then the associated consequences or impact to the risk assessing agent as a result of their variation should then be ascertained. We will discuss the steps to achieve this in the next sub-section.

6.10 Ascertaining the Resources to Be Kept at Stake Curve (RKC) in the Business Activity

As with dependent events, the steps to determine the extra financial resources to be kept at stake by the risk assessing agent due to variations in the occurrence of non-dependent events are as follows:

(a) Determine the RDC. This curve accurately represents the value of financial resources at stake that will be affected due to variations associated with the occurrence of non-dependent event/s.
(b) Determine the impact of variations associated with the occurrence of non-dependent event/s on the financial resources at stake (determined in step (a)).
(c) Determine the extra financial resources to be kept at stake in the business activity due to variations associated with the occurrence of non-dependent event/s.

We will explain each of the abovementioned steps in detail.

6.10.1 Ascertaining the Resources Decided at Stake Curve (RDC) in the Business Activity

The aim of this step is to accurately determine the value of financial resources decided to be kept at stake, which will be affected by variations associated with the occurrence of non-dependent events. To do this, the following steps should be carried out:

(i) Determine those time slots from the post-interaction start time phase of the business activity on which the non-dependent event may have an effect.
(ii) Determine the investment pattern of financial resources that will be affected by the non-dependent event in those time slots.
(iii) Plot a probabilistic curve from the analysis conducted in the previous step to represent the probability of an amount from the invested financial resources at stake in the time slots of the occurrence of the non-dependent event.

The risk assessing agent, when determining the VC, identifies the time slots from the post-interaction start time phase in which the non-dependent events may have an effect on its completion. While determining the impact due to variations associated with the occurrence of those events, he should consider only those time slots from the post-interaction start time phase and determine the variability in the investment of his financial resources (similar to scenarios 6.1 and 6.2 in Section 6.3). To accurately determine the resources at stake in those time slots, the risk assessing agent should plot the probabilistic curve known as the *Resources Decided at Stake Curve (RDC)*. The RDC is a probabilistic curve that shows the amount of financial resources at stake from the net financial resources decided in the expectations only in that time slot in which the non-dependent event will have an impact on the successful completion of the business activity. There are two possible scenarios:

Scenario 6.9: The non-dependent event that may have an effect on the successful completion of the interaction occurs in only one time slot of the post-interaction start time phase

If the non-dependent event occurs in only one time slot of the post-interaction start time phase (as mentioned in case 6.7.1), then the risk assessing agent should determine the investment pattern of his financial resources in that time slot and plot the RDC.

Scenario 6.10: The non-dependent event that may have an effect on the successful completion of the interaction occurs in more than one time slot of the post-interaction start time phase

If the non-dependent event occurs in more than one time slot of the post-interaction start time phase, as discussed in case 6.7.2, then the risk assessing agent should determine the investment pattern of his financial resources in those time slots and plot the RDC.

To illustrate with an example, let us consider the previously discussed interaction scenario in Section 4.2.2.1 of agent 'A' forming a business association with a logistics company (agent 'B'). The resources decided to be at stake in the expectations are 10,000 euros with an investment pattern as mentioned in Scenario 6.1. From the expectations formed, the risk assessing agent in his business association has identified one event as the non-dependent event (variation in the exchange rate of Australian dollar with the European euro) which, should it occur, may have an impact in the last time slot of the post-interaction start time phase. The financial resources at stake on which the non-dependent event may have an impact is 3,000 euros which may occur only in time slot 't5' of the time space. The investment pattern of the financial resources to be affected and the RDC corresponding to this is as shown in Figures 6.10 and 6.11, respectively.

In contrast to the AIC shown in Figure 6.6, it can be seen that the RDC is plotted by taking into consideration only time slot t5 and not all the time slots in the post-interaction start time phase. This is because the non-dependent event may impact the business activity only in that time slot and the financial resources on which it may have an impact are invested only in that time slot. So, the RDC which determines the probability of an amount being at stake from the net resources on which the non-dependent event may impact should be plotted only by taking into consideration that time slot. If the non-dependent event may have an impact on more than one time slot of the post-interaction time phase, then the financial resources on which it may have an impact in those time slots should be determined and then the RDC should be plotted by taking into consideration those time slots.

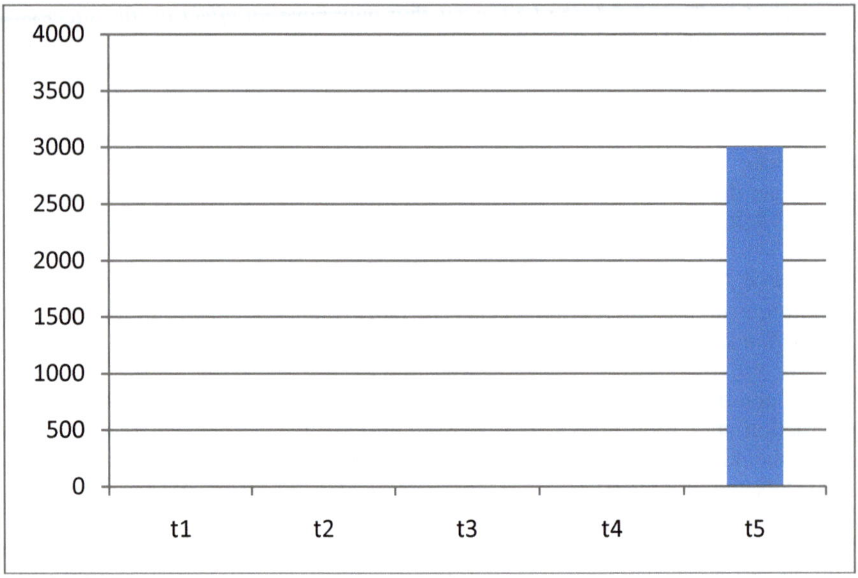

Fig. 6.10 The financial resources that will be affected by the non-dependent event in the time slot of the business activity

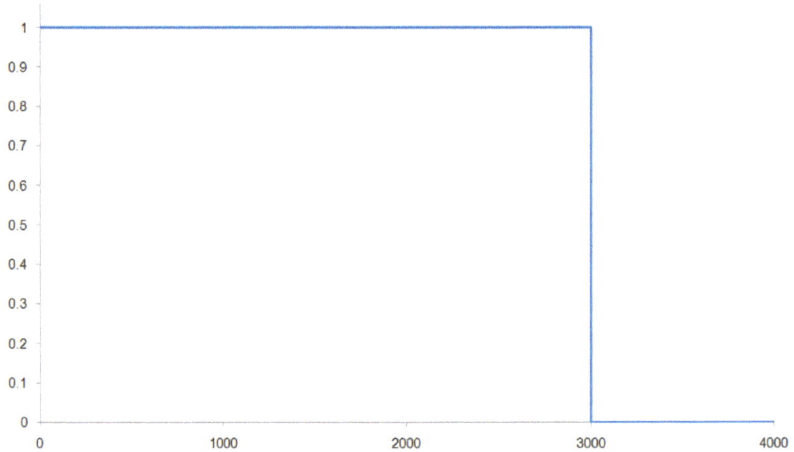

Fig. 6.11 The RDC according to the investment pattern shown in Figure 6.10

6.10.2 Determining the Financial Resources to Be kept at Stake due to the Non-occurrence of Non-dependent Events according to the Expectations

Once the RDC and VC in the business activity have been determined, then the risk assessing agent can ascertain the RKC in the business activity. To do this, we utilize the convolution operator discussed in Section 6.8.1 to plot the RKC. Mathematically, RKC can be determined as:

$$RKC = VC \oplus RDC \tag{6.11}$$

Before carrying out the convolution process, the VC (shown in Figure 6.9) should be normalized so that it represents only positive variations in the occurrence of a non-dependent event. This is because positive variations compared to negative variations in the occurrence of these events will result in the risk assessing agent keeping extra financial resources at stake in the business interaction which may contribute towards the occurrence of financial loss in the business activity, which is our focus of interest.

Extending the interaction scenario mentioned in the previous step, according to the expectations, the risk assessing agent 'A' has to pay 3000 euros to the logistic company upon receiving the goods. But, by considering the uncertainty in the occurrence of the non-dependent event, let us consider that the RKC which represents the financial resources to be kept at stake is as shown in Figure 6.12. The x-axis of Figure 6.12 represents the net worth of financial resources to be at stake in the business activity due to the variation in the occurrence of non-dependent events and the y-axis represents the probability of those resources to be at stake.

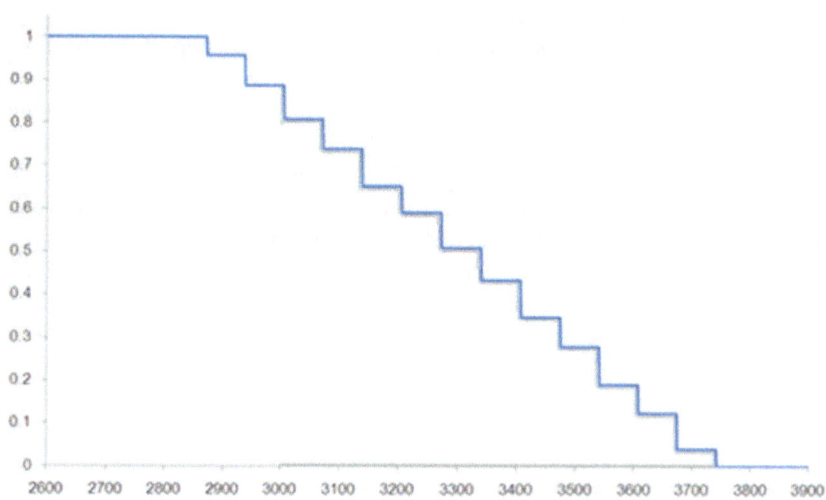

Fig. 6.12 The Resources to be Kept at Stake Curve (RKC)

Similar to the difference between the AIC and FAIC, the RKC as opposed to the RDC may be inflated and the level of inflation may represent the additional financial resources along with their probability of occurrence that the risk assessing agent may have to keep at stake in the particular time slot/s. If there is more than one non-dependent event in the post-interaction start time phase of the business activity, then the RKC for each should be determined.

After the RKC has been determined, the extra financial resources to be kept at stake due to uncertainty in the occurrence of non-dependent event/s, according to the expectations, should be determined. In other words, the extra financial resources to be kept at stake are the unserved investment by the risk assessing agent due to variations in the occurrence of non-dependent event/s from what was decided in the expectations. To calculate this amount, the *maximum decided resources (MDC)* should be plotted on the RKC. MDC represents the financial amount which the risk assessing agent had expected to invest in that particular assessment criterion in the expectations. Any additional resources invested after this point on the RKC represents the extra financial resources that the risk assessing agent may have to invest and keep at stake due to variations in the occurrence of the non-dependent event/s from what was decided in the expectations.

To illustrate with an example, let us consider the interaction scenario between agent 'A' and agent 'B' from the previous section. To determine the extra financial resources to be kept at stake, the risk assessing agent 'A' should plot the MDC on the abscissa of the RKC, which is represented by point 'w' in Figure 6.13. The resources beyond this point (represented by the shaded part of Figure 6.13) are the extra financial investments that the risk assessing agent may have to keep at stake due to the uncertainty in the occurrence of non-dependent events, as opposed to what was decided in the expectations.

If there is more than one non-dependent event in a business association, then the risk assessing agent should repeat the abovementioned steps from Sections 6.10.1 to 6.10.2 and determine the financial resources that it may have to keep at stake from the MDC for each event. It can then utilize the analysis to determine the financial risk in forming an interaction with the risk assessed agent. However, before determining the financial risk, the risk assessing agent should combine the shaded part of the RKC after its MDC for each non-dependent event shown in Figure 6.13 with the FAIC of the interaction shown in Figure 6.8 to ascertain the total financial resources that it may need to keep at stake in his business interaction with the risk assessed agent. In the next section, we present our approach to determine the total financial resources needed to be kept at stake in the business association.

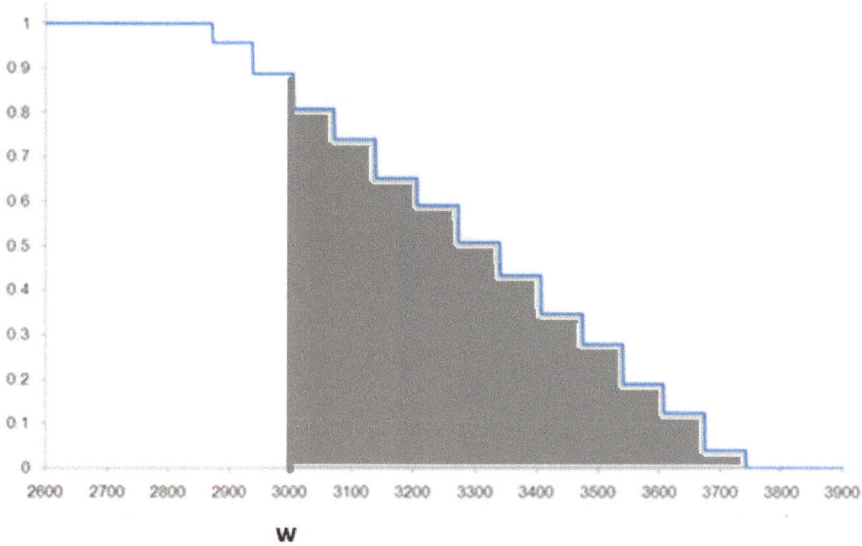

Fig. 6.13 The RKC showing the resources to be kept at stake due to variations in the occurrence of a non-dependent event from that decided in the expectations

6.11 Ascertaining the Total Factual Amount Invested Curve (TFAIC) of the Business Activity

The total financial resources to be kept at stake in the business activity by considering the uncertainty in the occurrence of both dependent and non-dependent events can be determined by a two-step process which is described in the following sections,

6.11.1 Ascertaining the Extra Investment Curve (EIC) of the Business Activity

The EIC represents the extra financial resources that the risk assessing agent has to keep at stake in the business activity due to variations in the occurrence of non-dependent event/s. The EIC of the business activity is determined by considering the shaded part of the RKC. To accurately determine the additional financial resources needed to be kept at stake in the business activity due to variations in the occurrence of non-dependent event/s, the part of the RKC after its MDC should be normalized to achieve a probability density function that satisfies the following condition:

$$\sum_{i=1}^{n} p(i) = 1 \tag{6.12}$$

where:

'i' represents the different levels of financial resources,
'n' represents the number of financial levels after the MDC.

To illustrate with an example, let us consider the RKC as shown in Figure 6.13 and the MDC of 3000. The resulting EIC from the non-dependent event/s is shown in Figure 6.14.

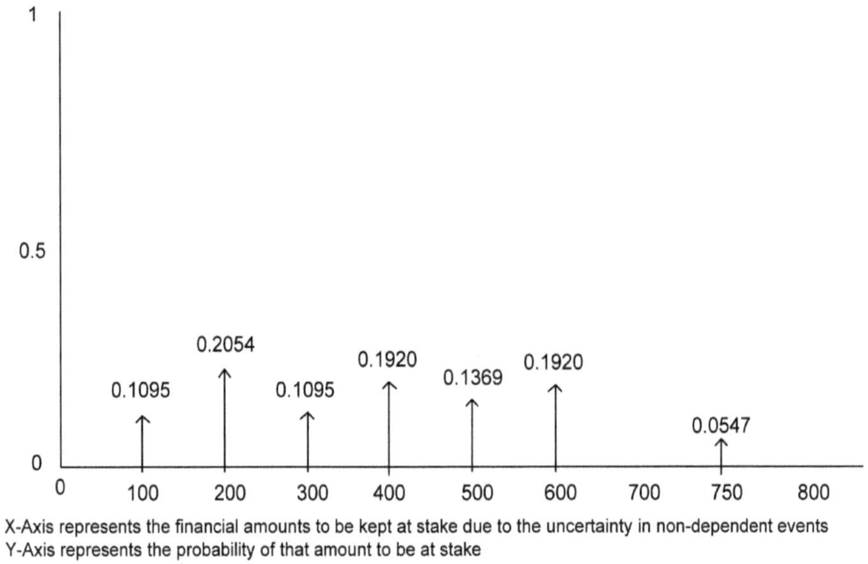

Fig. 6.14 The EIC from the RKC of Figure 6.13

6.11.2 Combining the EIC with the FAIC to Ascertain the TFAIC of the Business Activity

Once the EIC of the business activity has been determined, it should then be combined with the FAIC of the business interaction to determine the *TFAIC*. Mathematically, the process of determining the TFAIC of a business association by using the convolution operator is determined as:

$$\text{TFAIC} = \text{FAIC} \oplus \text{EIC}_n \tag{6.13}$$

where,

n represents the number of non-dependent events in the business activity.

Extending the interaction scenario between agent 'A' and agent 'B' from the previous section, the Total Factual Investment Curve (TFAIC) of the business activity by convolving the FAIC of Figure 6.8 with the EIC of Figure 6.14 is represented in Figure 6.15. The x-axis of Figure 6.15 represents the net financial resources to be kept at stake in the business activity and y-axis represents the probability of those resources to be at stake.

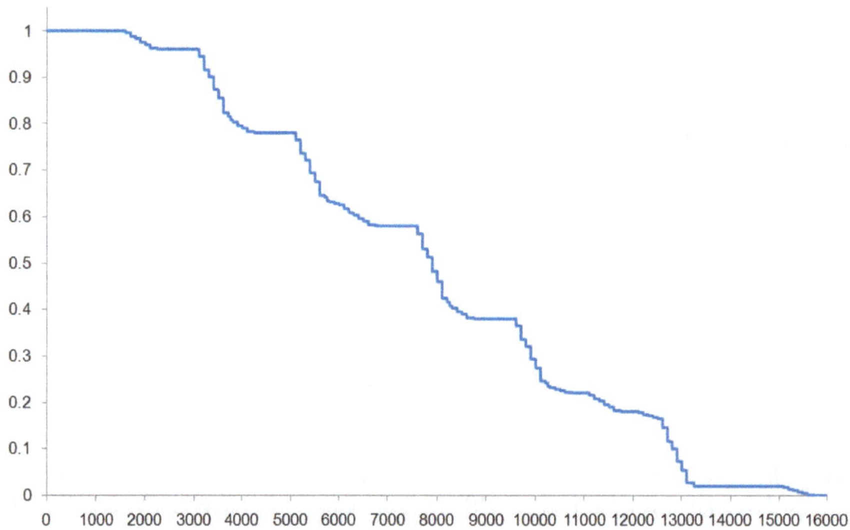

Fig. 6.15 The TFAIC in agent A's business activity with agent 'B'

It can be seen that the TFAIC for the business activity is inflated compared with the FAIC of Figure 6.8 due to the extra financial resources needed to be kept at stake as a result of variations in the non-dependent events. The higher the levels of resources and probability of these resources needed to be at stake in the EIC, the greater the inflation will be in the TFAIC compared to the FAIC. If there is more than one non-dependent event, then the risk assessing agent should determine the TFAIC by convolving the EIC of each non-dependent event with the FAIC. It can then utilize the TFAIC to ascertain its financial risk in forming a business association with the risk assessed agent. In the next section, we present our approach to determine the financial risk to the risk assessing agent in forming a business activity with a risk assessed agent.

6.12 Ascertaining the Financial Risk in Forming a Business Activity

Once the risk assessing agent ascertains the financial resources that he needs to keep at stake in the business activity, we propose that the following factors should be determined to ascertain the financial risk in the business activity:

- the probability that he will not achieve the full benefit of the resources that he invests;
- the levels of un-served investment or the amount of extra resources that has to be kept at stake in the business activity as opposed to what was decided in the expectations.

These factors will be ascertained on the TFAIC by plotting the *Maximum Investment Capacity (MIC)* of the risk assessing agent and determining the Loss of Investment Probability (LOIP) of the business activity.

6.12.1 Maximum Investment Capacity (MIC) of the Risk Assessing Agent

The MIC of the risk assessing agent represents the maximum amount of resources it can invest while forming a business association. This is usually the amount of resources decided while forming the expectations.

6.12.2 Loss of Investment Probability (LOIP) in the Business Activity

When the MIC of the business activity is plotted on the TFAIC, it represents the *Loss of Investment Probability (LOIP)* to the risk assessing agent. The LOIP of the business activity gives the probability of the risk assessing agent not achieving the full benefit of the resources that he invests while interacting with a risk assessed agent. This may be due to:

a) the risk assessed agent not meeting the risk assessing agent's expectations due to his performance risk; and/or
b) variations in the occurrence of non-dependent events in the business activity.

The LOIP index to the risk assessing agent in the business activity can be determined by ascertaining the amount of additional financial resources to be kept at stake, after the point of his maximum investment capacity. The LOIP index of an interaction is the ordinate of the TFAIC corresponding to the amount on the abscissa immediately after the risk assessing agent's MIC. By definition of TFAIC, this ordinate is the probability of the corresponding amount needed to be kept at stake, but this amount and amounts beyond this point will not be kept at stake by the risk assessing agent as they are outside his maximum investment capacity. So the Loss of Investment Probability (LOIP) to the risk assessing agent in the business activity is determined by:

$$\text{LOIP} = \text{TFAIC}\,(w) \qquad (6.14)$$

where:

w = the point immediately after the MIC of the risk assessing agent.

Continuing the interaction scenario between agent 'A' and agent 'B' from the previous section, agent 'A' expects to achieve all of his expectations in interacting with the risk assessed agent 'B' to his MIC of 10,000 euros. But there is the possibility of this not happening due to the performance risk of agent 'B' and variations in the occurrence of non-dependent events which results in having the TFAIC of the business interaction as shown in Figure 6.15. As a result, the LOIP to the risk assessing agent 'A' is the ordinate of the TFAIC at point 10,000 as shown in Figure 6.16.

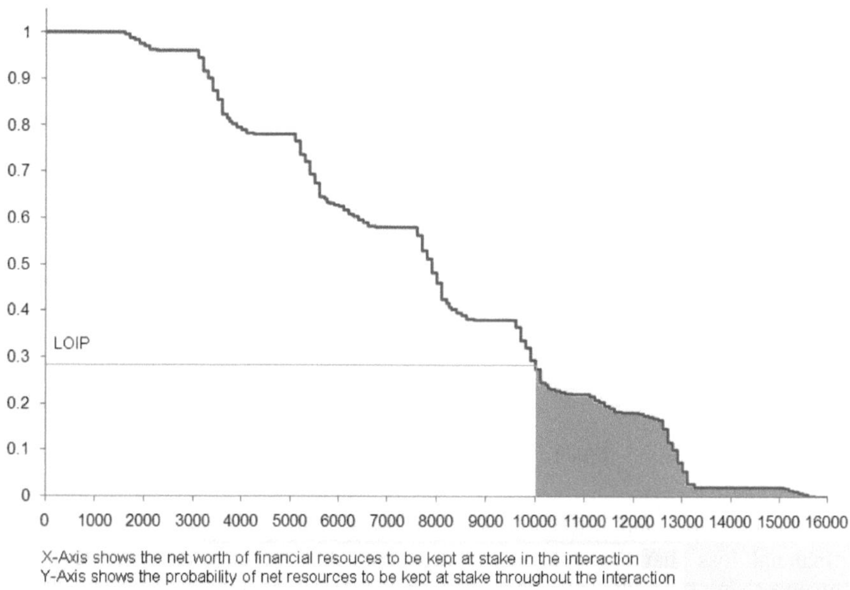

Fig. 6.16 Determining the Financial Risk in the Business Activity

Once the LOIP of the business activity has been determined, the financial risk to the risk assessing agent in forming a business activity with a risk assessed agent can then be ascertained by determining the LC of the business activity.

6.12.3 Ascertaining the Loss Curve (LC) of the Business Activity

The shaded part of Figure 6.16 shows the *Possible Consequences of Failure (PCF)* which represents the additional resources needed to be kept at stake which are beyond his maximum investment capacity. In other words, these are the un-served investments which theoretically are beyond the maximum investment capacity of the risk assessing agent. To determine the different levels of net financial loss that could be experienced in the business activity, the risk assessing agent should plot the cumulative probability function of the un-served amounts on the TFAIC, as shown in Figure 6.17. We term the cumulative probability function of the un-served amounts in the business activity the 'Loss Curve' (LC) [2].

The x-axis of Figure 6.17 represents the different level/s of financial amounts and y-axis represents the probability of experiencing a loss of that amount in the business activity.

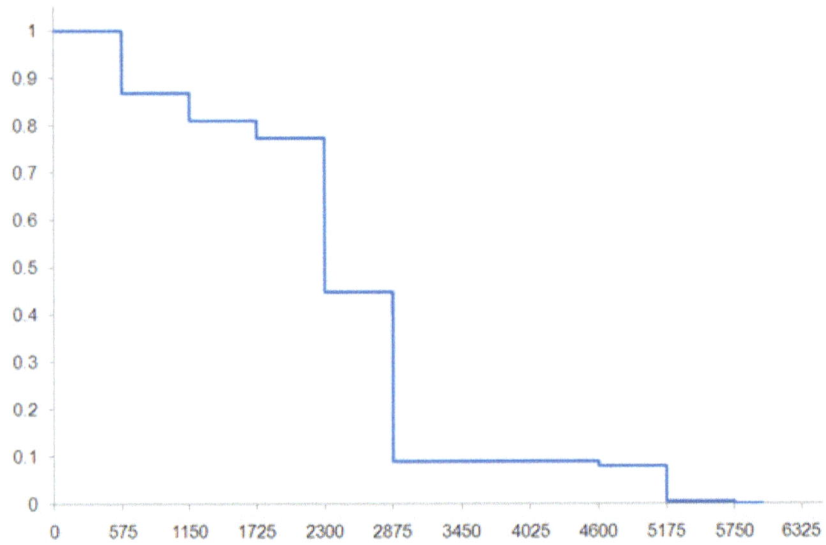

Fig. 6.17 The LC of the Business Activity

The LC represents a detailed and informed analysis of the different levels of financial loss that the risk assessing agent 'A' can experience in forming a business activity with risk assessed agent 'B' by taking into account the different types of uncertainties associated with his interaction. The risk assessing agent, after ascertaining the performance and financial risk in interacting with a risk assessed agent, can then determine the magnitude of perceived transactional risk in forming a business activity with that agent. We will discuss this in the next chapter.

6.13 Recapitulation

In this chapter, we proposed a methodology by which the risk assessing agent can ascertain the financial risk in forming a business interaction with the risk assessed agent. We began by firstly discussing the importance of identifying the object of analysis at stake, according to the outcomes to be achieved in the business activity. We then differentiated between the terms resources *invested* and resources *at stake* and then identified the different types of events that have to be considered to undertake financial risk assessment in business activities. We then proposed techniques by which the risk assessing agent can determine the financial risk by considering the uncertainties in the occurrence of dependent and non-dependent events. Such an analysis to quantify the level of financial loss in a business activity is novel and has not been discussed previously in the literature. Approaches have been proposed in the literature which determines the level of loss as a crisp financial amount. However, as the business association will be carried out at a future point in time when nothing is certain, it is quite difficult to accurately determine beforehand and be certain of the amount of the crisp financial loss that a risk assessing agent may experience in his business interaction. Our proposed approach overcomes this problem and proposes a method whereby the different amounts of financial loss, along with the probability of experiencing these, can be calculated by taking into consideration the different types of uncertainties in the business activity. In the next chapter, we propose a methodology by which the risk assessing agent, based on the analysis of performance risk and financial risk, can determine the perceived transactional risk in a business activity.

References

1. Hussain, O.K., Dillon, T., Hussain, F., Chang, E.: Probabilistic Assessment of Loss in Revenue Generation in Demand-Driven Production. Journal of Intelligent Manufacturing (2011), doi:10.1007/s10845-011-0518-4
2. Hussain, O.K., Dillon, T., Hussain, F., Chang, E.: Probabilistic Assessment of Financial Risk in E-Business Association. Simulation Modelling Practice and Theory 19, 704–717 (2011)
3. Sanabria, L.A., Dillon, T.S.: Stochastic power flow using cumulants and Von Mises functions. Electrical Power and Energy Systems 8, 47–80 (1986)
4. Sanabria, L.A., Dillon, T.S.: An Error Correction Algorithm for Stochastic Production Costing. IEEE Transactions on Power Systems 3, 94–100 (1988)
5. Zhang, P., Lee, S.T.: Probabilistic Load Flow Computation Using the Method of Combined Cumulants and Gram-Charlier Expansion. IEEE Transactions on Power Systems 19, 676–682 (2004)
6. Kendall, M.G., Stuart, A.: The advanced theory of statistics. Griffin (1968)
7. Gnedenko, B.V.: Theory of Probability. Chelsea Publishing Company, New York (1966)
8. Evans, J.R., Olson, D.L.: Statistics, Data Analysis, and Decision Modeling. Prentice Hall (2002)

Chapter 7
Risk Assessment Phase: Transactional Risk Assessment in Business Activities

7.1 Introduction

In the previous two chapters, we proposed an approach by which each subcategory of perceived transactional risk (transactional risk), namely performance risk and financial risk in a business activity, can be determined by considering its context, assessment criteria and dynamic specific characteristics. The next step in the process of transactional risk assessment is to assimilate the determined analysis of the subcategories and ascertain the degree of risk. The determined risk level represents the combined effect of the level of failure and the associated consequences in forming a business association with the risk assessed agent. This will form important input into the next two phases, namely transactional risk evaluation and risk management (or risk treatment), which are important steps in the process of risk analysis. In this chapter, we present an approach by which the risk assessing agent can determine the degree of transactional risk in forming a business activity with a risk assessed agent.

7.2 Variability and Uncertainty to Be Captured during Transactional Risk Assessment

Before the level/s of transactional risk in a business activity can be ascertained, it is important to identify the variability and uncertainty of its occurrence during its assessment. To illustrate with an example, let us consider a business activity scenario between risk assessing agent 'A' and risk assessed agent 'B'. Let us consider that a scale of 0-100% is used to represent the transactional risk and using this scale, risk assessing agent 'A' determines the level of transactional risk in forming a business association with risk assessed agent 'B' is 80%'. As shown in Figure 7.1, this leads to an understanding that:

(a) on a scale of 0-100, the severity of transactional risk in the business activity is 80% (a crisp level), and
(b) on a scale of 0-1, the likelihood of that risk occurring is 1.

O.K. Hussain et al.: Risk Assessment & Management in the Networked Economy, SCI 412, pp. 187–215.
springerlink.com

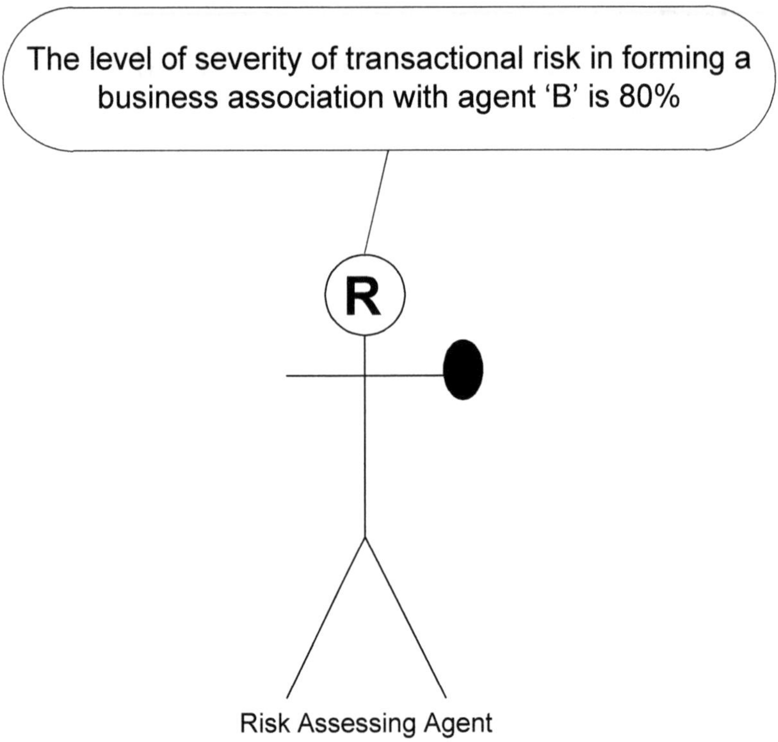

Fig. 7.1 An interaction scenario between agent 'A' and agent 'B'

One of the scenarios in which such a uni-level of transactional risk may be possible is when the business association of agent 'A' with agent 'B' is limited to the current period of time, and the risk assessing agent has his own past interaction history with the risk assessed agent, which leads to the crisp or uni-level occurrence in both the sub-categories of transactional risk. However, having such a uni-level of transactional risk may not be possible or realistic in every interaction scenario. Some such scenarios are as follows:

(a) if the business activity of agent 'A' with agent 'B' extends to a point of time in the future, and
(b) agent 'A' solicits recommendations for agent 'B' in the pre-interaction start time phases. As discussed in Chapter 5, assimilating them may introduce variability or multi-levels of performance risk in the pre- and post-interaction start time slots; which will lead to multi levels of financial risk as discussed in Chapter 6 (for example, as shown in Figures 5.20 and 6.16 respectively).

In such scenarios, it may not be realistic for the risk assessing agent 'A' to conclude with a uni-level or crisp representation of the transactional risk that could be present throughout the business interaction time period. Rather, he needs to capture the *uncertainty* represented by the different level/s of severity of performance risk and financial risk and map it while determining the transactional risk. There are two types of uncertainty when assessing transactional risk in such scenarios. They are:

(a) determining which level/s of severity of transactional risk that might occur in the business activity; and
(b) determining the likelihood of occurrence of those level/s.

The first type of uncertainty aims to identify the different level/s of severity in the occurrence of transactional risk in the business activity on the given scale. For example as shown in Figure 7.2, the risk assessing agent determines the level/s of

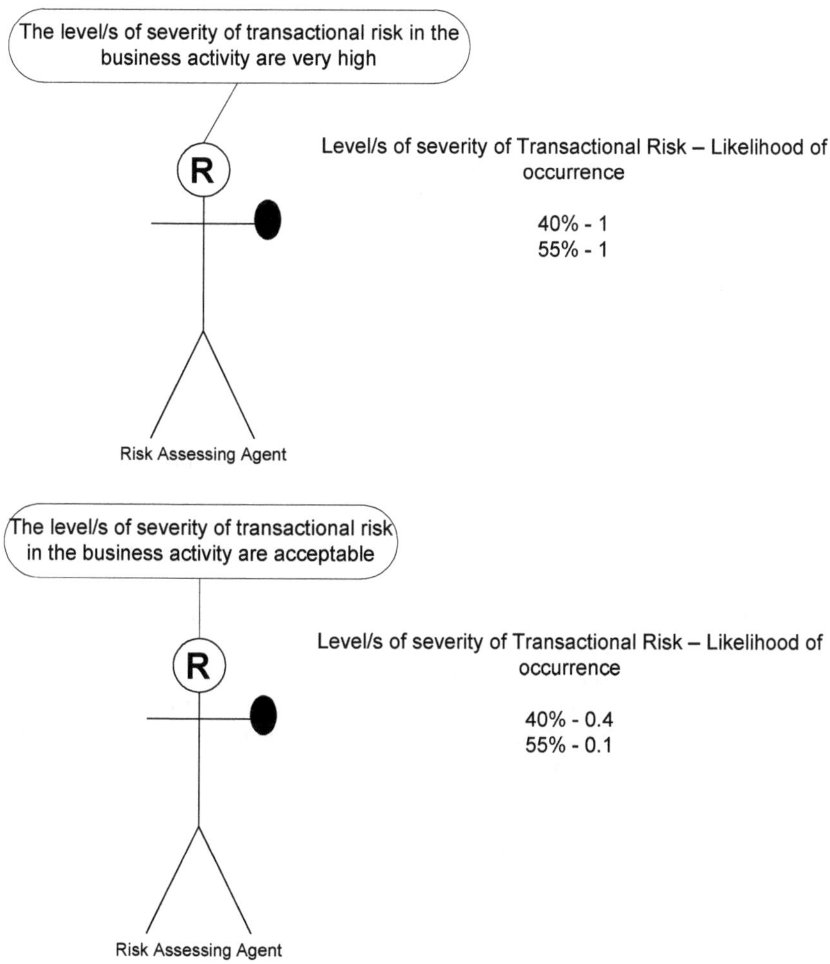

Fig. 7.2 Effect of the variation in the likelihood of occurrence of the severity levels of transactional risk on the next phases of risk analysis

severity of transactional risk in the business activity as 40% and 55%. The second type of uncertainty captures the likelihood of occurrence of these level/s. This will have an important effect in the understanding of transactional risk. For example, if the likelihood of occurrence of the levels of transactional risk (40% and 55%) is 1 (on a scale of 0 to 1), then we would regard the transactional risk as high. However it is not necessary that the likelihood of occurrence of these levels is 1. Rather, it could be that the likelihood of occurrence is 0.4 and 0.1, respectively as shown in Figure 7.2. If this is the case, then we would regard the level of transactional risk not as high as considered earlier. This is captured by the second type of uncertainty that aims to identify the likelihood of occurrence of those levels. As shown in Figure 7.2, both these types of uncertainty represent different analyses that are important to understand as they will have an important effect in defining the process of risk evaluation and risk management.

So, during transactional risk assessment, the uncertainty which is present in the risk assessing agent's mind at this stage can be classified into two types, as discussed in the next sub-sections.

7.2.1 *Uncertainty in Identifying the Severity of Transactional Risk*

In the context of this discussion, this type of uncertainty in the risk assessing agent's mind represents the fuzziness in identifying the level/s of severity of transactional risk present in interacting with the other agent. For example, what level/s of severity in transactional risk is present in agent 'A's business activity with agent 'B'?

7.2.2 *Uncertainty in Identifying the Likelihood of Transactional Risk*

This type of uncertainty represents the fuzziness in identifying the degree or likelihood to which those level/s of severity will occur in the interaction. For example, from the identified level/s of transactional risk above, what is their likelihood of occurrence in the business activity?

Both these types of uncertainties have to be understood and measured during transactional risk assessment. However, before this can be done, a commonly agreed scale needs to be defined so that such representations can be made. We discuss this further in the next section.

7.3 Scale to Measure Transactional Risk in Business Activities

As discussed in Chapter 3, the aim of the risk assessment step is for the risk assessing agent to identify and measure the different level/s of transactional risk in forming a business association with a risk assessed agent. However, before this can done, one of the pre-requisites is for the risk assessing agent to have a range over which the possible level/s of transactional risk in a business activity can be determined. As discussed in Chapter 5 and represented in Figures 5.2 – 5.4, failure to have a standardized representation would lead a risk assessing agent using his own range for assessment and representations. This may create ambiguities or irregularities:

(a) in the utilization of the determined level/s of transactional risk by the risk assessing agent during the process of risk evaluation and management, or
(b) in making recommendations to other agents during the process of risk communication, or
(c) in the other agents correct understanding of the severity of transactional risk as communicated by the risk assessing agent, or
(d) if the risk assessing agent utilizes the current interaction as a reference at a future period of time for a different interaction.

Furthermore, as discussed in the previous section, the range over which the transactional risk of the business activity is being determined should be identified or chosen such that it captures the two different types of uncertainty. To have such a range, we propose that a numeric scale of 0-100 is used to identify and measure the different level/s of severity of transactional risk. To represent the likelihood of the occurrence of those levels, a scale of 0-1 is used. Each value on these scales has its own expression, which appropriately identifies and measures the severity of occurrence as well as the likelihood of occurrence of the level/s of transactional risk.

In the next section we discuss the importance of defining and associating semantics with the transactional risk representation.

7.4 Importance of Semantics in Transactional Risk Representation

As mentioned earlier, the broad aim of the risk assessment step is to determine the level/s of transactional risk in forming a business association. However, it is important that the determined level/s of transactional risk are represented in a way that can be understood by the risk assessing agent. Failure to do this may result in ambiguities or irregularities explained in the previous section and in the risk assessing agent misunderstanding what they mean and represent, despite having the level/s of transactional risk. If this is the case, then the whole purpose of undertaking risk analysis is lost.

For example, let us consider agent 'A' forming a business association with the agent 'B' in forming a business association, as shown in Figure 7.3. On the scale of 0-100, it is a commonly understood notion that the closer the value towards 100, the higher is the representation of the factor being represented. By such a representation, even though agent 'A' knows the level/s of severity of transactional risk in forming an interaction with agent 'B', that representation would be fuzzy and vary according to agent 'A's interpretation of those levels, as shown in Figure 7.4.

In other words, the question that arises is what does agent 'A' make of those levels of severity? Does he consider the degree of transactional risk to be 'low' or 'moderate' or 'high' or in other terms 'acceptable' or 'un-acceptable'? To address

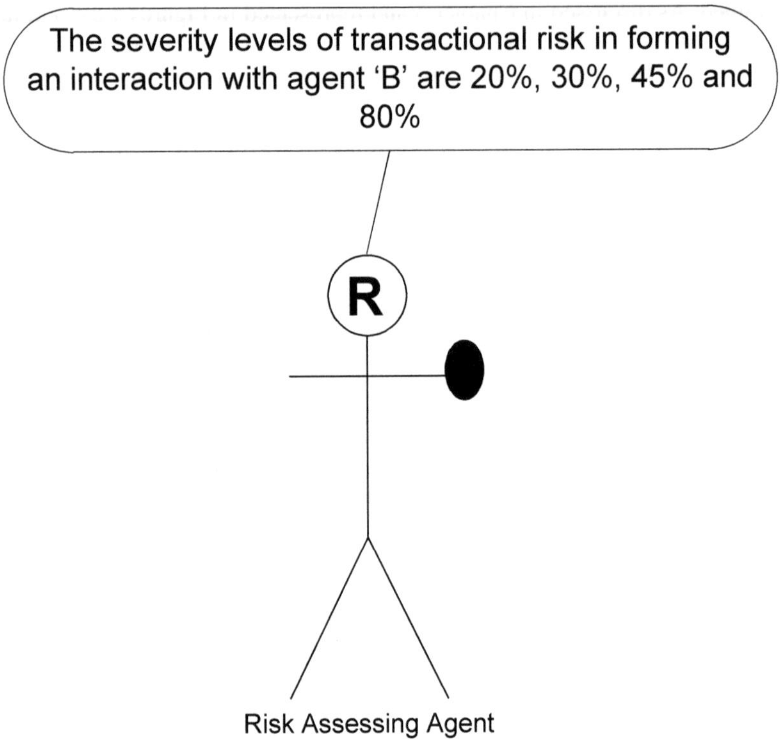

Fig. 7.3 Level/s of severity of transactional risk in the interaction between agent 'A' and agent 'B'

such fuzziness in the understanding of the level/s of transactional risk, a more concrete and standardized understanding of its representation; apart from having it on a commonly understood scale (0-100) is needed. This is necessary to assess the severity of a risk level, which is important in the phases of risk evaluation and risk management, as these steps mainly depend on identifying and addressing those level/s of transactional risk, from the determined ones, which are above the acceptable risk tolerance levels of a risk assessing agent.

So, to address scenarios such as these, apart from having a scale to determine the level/s of transactional risk in forming a business interaction, it is also important to transform the analysis from the scale to a representation that can be commonly understood and which can be utilized for further analysis in the risk assessment process. This can be done by using the concept of *semantics*. Meadow [1] presents the four different stages in the conversion of data to text which can be widely understood. We give a brief discussion of each stage in the context of the discussion of this book.

The first stage is *data* which the risk assessing agent gathers prior to the assessment of each sub-category of transactional risk. When data is put in the

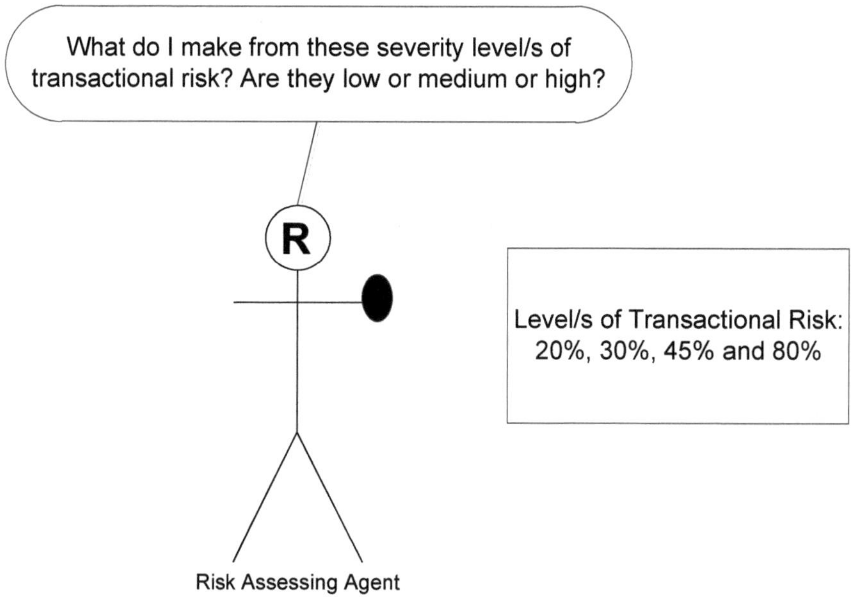

Fig. 7.4 Understanding the level/s of severity of transactional risk in the business activity

context of the analysis of each sub-category of transactional risk, it then becomes *information*. When information is utilized to determine first the sub-categories of transactional risk and later the transactional risk itself, then it becomes *knowledge*. But, for the knowledge to make sense, it should be associated with a *meaning* which is achieved by utilizing *semantics*. Semantics is the key for understanding ubiquitous information according to the desired way to carry out sophisticated tasks [2]. Associating semantics with a given piece of information will give it meaning, making it easier to understand, interpret and utilize it for further tasks.

So, by adding semantics to the determined level/s of transactional risk, the risk assessing agent will be able to assess the analysis better and to utilize it in an informed way in the next phases of risk evaluation and risk management. In the next section, we propose two different representations of transactional risk in order to understand it in clear and unambiguous terms.

7.5 Different Representations of Transactional Risk in Business Activities

To have a clear and unambiguous understanding of transactional risk in a business activity, we propose two representations as follows:

7.5.1 Numeric Representation of Transactional Risk

Numeric representation of transactional risk represents the determined level/s of severity of transactional risk on a numeric scale of 0-100 in forming a business activity. On this scale, the value 0 represents the lowest level of severity of transactional risk and a value of 100 represents the highest level of severity of transactional risk. To represent the likelihood of the occurrence of those levels of severity, a scale of 0-1 is used; with the value 0 representing the lowest likelihood of occurrence whereas the value of 1 represents the highest likelihood of occurrence. As mentioned earlier, each value on these scales has its own expression, which captures and represents the severity of the occurrence along with the likelihood of occurrence of those severities of transactional risk. However these values do not have semantics associated with them. This is addressed by having a linguistic representation of transactional risk.

7.5.2 Linguistic Representation of Transactional Risk

As mentioned earlier, even though the numeric level/s of transactional risk represents the severity of its occurrence, that representation can be fuzzy and vary according to each agent's interpretation. This does not help the risk assessing agent to ascertain and understand the accurate level of transactional risk in the next phases or over a period of time. This is addressed by the introduction of semantics and by determining the *linguistic representation of transactional risk*. Such representation of transactional risk defines semantic tags in the range of values between 0-100 and determines the level/s of severity of occurrence of transactional risk in those tags. This will play an important role in the risk evaluation and risk management phases.

By ascertaining both the numeric (on a scale of 0-100) and linguistic (on the defined semantic tags) level/s of severity of transactional risk, the risk assessing agent will have a clear picture of its magnitude, severity and likelihood of occurrence; and can assess the results in an informed way during the next phases of risk analysis. In the next section, we discuss the techniques we utilize to ascertain the numeric and linguistic representation of transactional risk.

7.6 Techniques for Determining the Numeric and Linguistic Representation of Transactional Risk

As mentioned in Section 7.2, it is important to capture the uncertainty that represents the variability in the occurrences of the different level/s of severity of transactional risk and the likelihood of occurrence of these level/s in the business activity. This applies while determining the two representations of transactional risk discussed above. In this section, we discuss the techniques which we utilize in our approach to identify such types of uncertainties while ascertaining the numeric and linguistic level/s of transactional risk.

7.6.1 Possibility Theory to Determine the Numeric Representation of Transactional Risk

Probability theory has been widely used in the literature in various applications to determine the chances of an event occurring over a given period of time [3]. However, apart from probability theory another theory, known as the possibility theory is better suited to determine and represent uncertainty of an event occurring at the future period of time. As the name implies, possibility theory determines the possibility of a variable occurring from its universe of discourse. In the given problem, each value from the defined universe of discourse is taken and the possibility of the occurrence of that value is determined. The idea of possibility theory was first coined by Lofti Zadeh in 1977 and later enhanced by Dubois and Prade. It has been used in the literature to estimate various complex uncertainties in real-world problems, for example terrorist risk [4]. In our approach, we utilize it to ascertain the possibility of occurrence of different level/s of severity of transactional risk over a period of time in the future.

Possibility theory determines the plausibility of each element in a range of 0-1 from the given set or universe of discourse occurring depending on the level/s of transactional risk present. Represented mathematically, the possibility distribution of a value v is as shown below:

$$\pi_{\mathrm{TR}}(v) \in [0, 1] \tag{7.1}$$

For any value v, if the possibility of occurrence has a value of 0, then this means that given the current inputs, there is no possibility of that value (in our case the level of severity of transactional risk) being present. When the possibility of occurrence of value v is different from 0, then this means that given the current inputs, the likelihood of that value (level of severity of transactional risk) being present in the business association is positive. For our problem, such characteristics of the possibility theory suits well, as by undertaking such an analysis:

(a) the horizon over which the transactional risk is being determined is widened, thus allowing the risk assessing agent to identify the different level/s of severity and the likelihood of transactional risk present in the business activity, and
(b) the possibility of the occurrence of those level/s of severity of transactional risk will be mutually exclusive from the other possible level/s of risk present in the business association.

So, in our approach, we utilize possibility theory to determine the numeric level/s of transactional risk in forming a business association. For further details on possibility theory or to gain a better understanding of possibility theory, readers are encouraged to refer to the work by Zadeh [5], Dubois and Prade [6, 7] and Zimmerman [8]. Further discussion on determining the numeric representation of transactional risk is in section 7.8.

7.6.2 Fuzzy Inference System to Determine the Linguistic Representation of Transactional Risk

To determine the linguistic level/s of transactional risk, an approach is needed which, based on the numeric inputs, computes the different level/s of severities of transactional risk as output on the defined semantic tags. There are different techniques to achieve this. One such approach is fuzzy inference system. Fuzzy systems are mathematical objects that model vagueness and uncertainty when the described phenomena do not have sharply defined boundaries. They were developed to incorporate the concept of partial truth characterized by the fuzziness of the data which yields a more accurate mathematical representation of the perception of truth than that of crisp sets [9]. They provide a precise approach for dealing with uncertain information by using a multi-valued logic derived from fuzzy sets. A fuzzy inference system models the vague inputs in terms of semantics and transforms them into a mathematical representation of the data to map to its output semantics. In our approach, we aim to utilize this property to ascertain the linguistic level/s of transactional risk. Further discussions on determining the linguistic representation of transactional risk is in Section 7.10.

In the next sections, we define the inputs to be considered and present the approaches to determine the numeric and linguistic level/s of transactional risk.

7.7 Inputs for Ascertaining Transactional Risk in a Business Activity

In Chapters 5 and 6, we explained in detail the subcategories of transactional risk and the various concepts associated with them, namely the FailureLevel Curve (FLC), Loss of Investment Probability (LOIP) and Possible Consequences of Failure (PCF). We explained that the impact of the FLC is determined on the Amount Invested Curve (AIC) to ascertain the Factual Amount Invested Curve (FAIC). Then, the non-dependent events and the variation in them are considered to ascertain the Total Factual Amount Invested Curve (TFAIC) of the business association. The Maximum Investment Capacity (MIC) of the risk assessing agent is then plotted to ascertain the Loss of Investment Probability (LOIP) and Possible Consequences of Failure (PCF) of the business activity. While ascertaining transactional risk in a business activity, we consider that the risk assessing agent considers only the PCF and LOIP of the business activity. This is to avoid having a double impact of the performance risk of the risk assessed agent (FLC) during the analysis, as it is considered during the financial risk analysis (to ascertain the TFAIC) in the business activity. So, the inputs and output while determining the numeric and linguistic level/s of transactional risk in forming a business association are as follows:

Inputs:

- Possible Consequences of Failure (PCF);
- Loss in Investment Probability (LOIP).

Output:

- Level/s of Transactional Risk (TR).

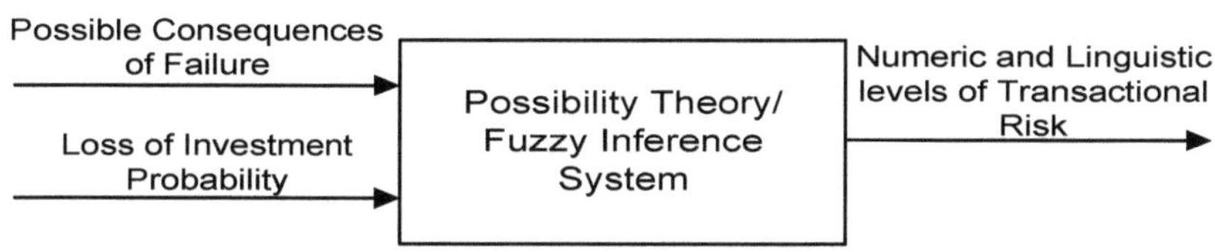

Fig. 7.5 The inputs to ascertain the numeric and linguistic level/s of Transactional Risk

The scale or the Universe of Discourse (UoD) over which the input variables and the output variable is defined by the following sets [10]:

- *Possible Consequences of Failure (PCF)* = {0, 1, 2, 3………. 100} where each element has a unit of %.
- *Loss of Investment Probability (LOIP)* = {0, 1, 2, 3…………..100} where each element has a unit of %.
- *Transactional Risk (TR)* = {0, 1, 2, 3……...100} where each element has a unit of %.

In the next sections, we utilize the inputs and output over the defined range and ascertain the numeric and linguistic level/s of transactional risk in forming a business association.

7.8 Steps to Ascertain the Numeric Level/s of Transactional Risk in a Business Activity

The series of steps to be followed while ascertaining the numeric level/s of transactional risk in a business activity are as follows:

7.8.1 Identify the Focal Elements of the Input Variables

Based on the determined LOIP and PCF in the business activity (from Chapter 6), the aim of this step is to identify the values from the UoD of the input variables which have a possibility of occurrence in the business activity. As mentioned in Dubious [7], the simplest form of possibility distribution of a variable x on the subset E on set u is:

$$\pi_x(u) = 1 \qquad \text{if } x \in E,$$

otherwise 0.

As shown in Figure 7.6, such a characteristic of possibility distribution will help in determining the range over which the inputs and transactional risk on the universe of discourse extend.

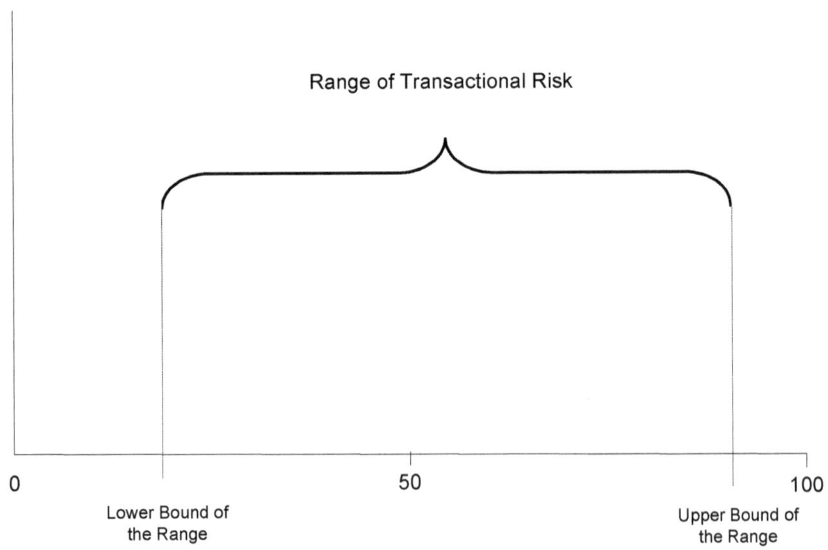

Fig. 7.6 Range or bounds over which transactional risk exists

To ascertain the range of transactional risk in our approach, the inputs of PCF and LOIP have to be transformed first to the possibility distribution over their universe of discourse. In other words, the likelihood of occurrence of each input element from its universe of discourse should be determined. This likelihood of occurrence of an element is termed the '*degree of evidence (DOE)*' of its outcome, represented by:

$$\text{DOE(X)} = \text{'m(A)'}$$

where 'A' is an element or unit from the universe of discourse of the input variable 'X'.

From the universe of discourse, those elements with a degree of evidence greater than zero are called the 'focal elements' for the particular input variable [4]. These elements represent the sets from the UoD for that variable upon which the evidence of occurrence focuses and which furthermore will be utilized from that input variable to determine the level/s of transactional risk in the business association. The degree of evidence of an element from the UoD should be in the interval between [0, 1] and the cumulative sum of the degree of evidence of all the focal elements from the UoD should satisfy the condition [11]:

$$\sum_{A \in X} m(A) = 1 \tag{7.2}$$

where:
'A' represents the focal elements belonging to the input variable 'X',
m(A) represents the degree of evidence of the focal element.

7.8.2 Determine the Possibility of Occurrence of Each Focal Element of the Input Variable

The aim of this step is to determine the possibility of occurrence of each focal element of the input variable. The possibility of an element 'A_i' of the input variable X can be determined from the focal elements of its UoD by [4, 11]:

$$\pi\ (A_i) = \sum_{j=1}^{n} m(Aj) \tag{7.3}$$

where:
$m(A_j)$ represents the degree of evidence of 'Aj' from the focal elements of the input variable from its UoD, and which have been ordered such that $i < j$ and $\pi(A_i) \geq \pi(A_j)$.

m(A) represents the degree of evidence that an element 'A' belongs exactly to 'X'. The *Belief* that an element belongs to 'X', Bel (A) represents the total evidence that the element belongs to 'X' as well as to any subset of 'X'. *Plausibility*, Pl (A) is the total evidence that the element belongs to 'X' as well as to any subset of 'X', plus additional evidence for sets that overlap with 'X'. Belief/Plausibility measures are referred to as Necessity (N)/Possibility (Π) measures, respectively [4]. Under such conditions, the possibility measures can be represented by a possibility distribution function as follows [4]:

$$\Pi\,(X) = \max\ \{\ \pi\,(a) \mid a \in X\} \tag{7.4}$$

where:
π (a) is the possibility of the element 'a'.

Equation 7.4 is repeated for each focal element 'X' of both the input variables to determine the possibility of its occurrence.

7.8.3 Identify the Focal Elements of the Output Variable

The aim of this step is to determine the focal elements of the output variable, namely the transactional risk in the business activity. There are different techniques that can be utilized for this process, but in our approach, we utilize the convolution operator. An important point to note here is that the process by which the convolution operation works in a possibility distribution is different from the one described in the previous chapter for probability distributions. The convolution of the focal elements from the UoD of the input variables (represented by variables *X* and *Y*) is done by taking their artesian products and is represented by [4]:

$$X \oplus Y = \{<x + y>: \text{where } x \in X \text{ and } y \in Y\} \tag{7.5}$$

where:
<x, y> denotes the tuple which represents the artesian product of the input focal elements from their UoD.

7.8.4 Determine the Possibility of Occurrence of Each Focal Element of the Output Variable

The possibility distribution that represents the likelihood of occurrence of the focal elements of the resultant output variable as the result of the convolution of the inputs variables is represented by [4]:

$$\pi(u) = \max\{\min[\pi_X(x), \pi_Y(y)]\} \tag{7.6}$$

where:
u is the focal element of the output function determined as the artesian product of the inputs f (x, y),
π (u) is the possibility of focal element ‘u’ from the output universe of discourse.

Following these steps, the risk assessing agent can ascertain the numeric level/s of transactional risk in a business activity. In the next section, we explain with an example, the process of ascertaining the numeric level/s of transactional risk in a business activity.

7.9 Ascertaining the Numeric Level/s of Transactional Risk in a Business Activity

Let us consider a business interaction scenario of risk assessing agent ‘A’ forming a business association with agent ‘B’ and the TFAIC of their business association, as shown in Figure 7.7. The first step in determining the numeric level/s of transactional risk is to identify the focal elements and possibility distribution of the input variables by using the steps discussed in the previous section. In the next sub-section, we propose the steps to be followed for transforming the input elements to the possibility distribution over their universe of discourse.

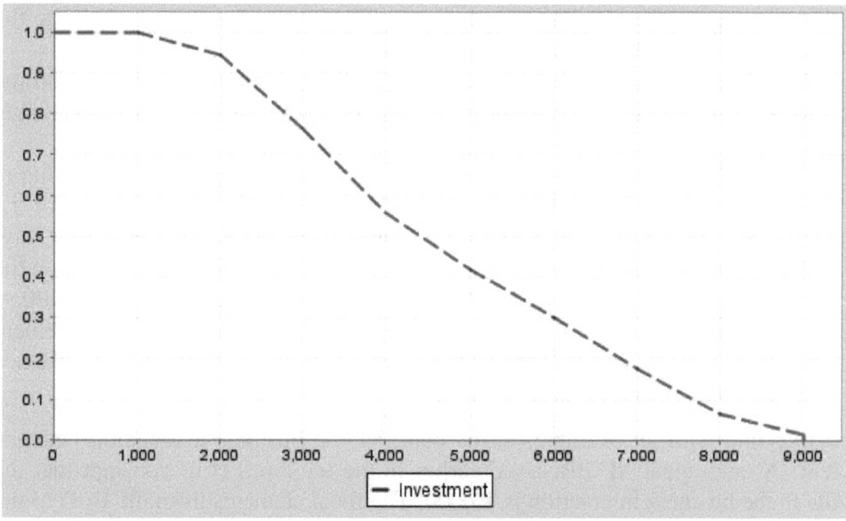

Fig. 7.7 The TFAIC of agent 'A's business association with agent 'B'

7.9.1 Steps to Transform the Input Elements to Possibility Distribution

As discussed in the previous chapter, the concepts of Loss of Investment Probability (LOIP) and the Possible Consequences of Failure (PCF) in forming a business association with a risk assessed agent are determined from the Total Factual Amount Invested Curve (TFAIC). To determine the focal elements and the possibility distribution of the PCF & LOIP in forming the business association, the various levels of extra financial resources or the levels of un-served investment which has to be kept at stake in forming the business association should be ascertained. The series of steps to achieve this are [10]:

- The Probability Mass Function (PMF) of the TFAIC in forming a business association with the risk assessed agent should be determined. The PMF represents the probability of each level of financial amount that the risk assessing agent has to keep at stake throughout the duration of his proposed business association with the risk assessed agent.
- The Maximum Investment Capacity (MIC) of the risk assessing agent (which is termed '*x*') should be determined.
- From point '*x*', the level/s or amount of extra financial resources that have to be kept at stake or the level/s of un-served investment/s in the business association should be determined.

- From the UoD of the PCF, the focal elements should be chosen according to the level/s of un-served investment from the MIC.
- Once the focal elements have been determined, their degree of evidence *m(A)* should be ascertained. This is done by taking into consideration the PMF of the particular financial amount from the TFAIC and then converting it to the possibility distribution, satisfying equation 7.2.
- Based on the degree of evidence calculated for each focal element from the UoD, the possibility distribution of the PCF can be determined by using equation 7.3 – 7.4.
- The LOIP of the interaction in contrast to the PCF is a single crisp value in the range of [0-1], which shows the ordinate on the TFAIC at the end of the risk assessing agent's maximum investment capacity in the business activity. But, in order to utilize a unified and comparable numerical scale for both the two inputs, the range of the LOIP is normalized in the range between 0-100.
- Hence, the ordinate of the TFAIC after point '*x*' is taken as the focal element from its UoD to represent the LOIP in the interaction. The degree of evidence of the focal element is taken as 1.

To illustrate with an example, let us consider the business association between agent 'A' and agent 'B' discussed earlier in the section. Let us consider that the MIC in the business interaction is $6,000. The focal elements from the UoD of the input variables PCF and LOIP are represented in Table 7.1 and Table 7.2 respectively.

Table 7.1 The focal elements and their degree of evidence for the input: PCF

Focal Elements of PCF	Degree of Evidence
16.66	0.61
33.32	0.311
49.98	0.079

Table 7.2 The focal elements and their degree of evidence for the input: LOIP

Focal Elements of LOIP	Degree of Evidence
30.2	1
40	0

Once the focal elements and their degree of evidence for each input variable have been determined, then equations 7.5 and 7.6 can be utilized to determine the focal elements of transactional risk and the possibility of occurrence of those elements. Continuing the interaction scenario between agent 'A' and agent 'B', the degree of evidence for the focal elements representing the severity of transactional risk in the interaction determined by using equation 7.6 is shown in Table 7.3 and the possibility distribution is shown in Figure 7.8.

Table 7.3 The focal elements and their possibilities for the output variable: Transactional Risk

Focal Elements of TR	Possibility of occurrence the Focal element
46.86	1
56.66	0
63.54	0.39
73.32	0
80.18	0.079
89.98	0

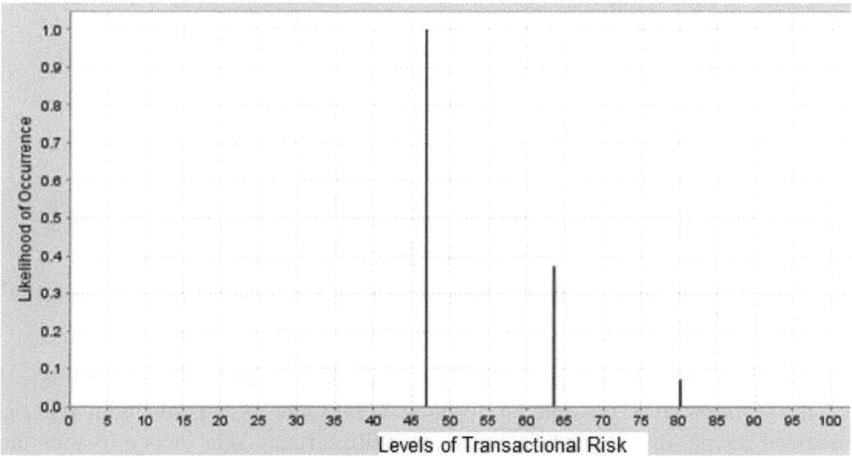

Fig. 7.8 Possibility distribution of Transactional Risk in agent 'A's business activity with agent 'B'

Some interesting observations from the above analysis are:

1. Only those elements from the UoD of an input variable which have a positive degree of evidence (i.e. which are the focal element from that input variable) have an effect on the output set during convolution. For example, we had taken the element 40 from the UoD of the input variable LOIP that has a degree of evidence of 0 and this did not have any effect while determining the level/s of severity of transactional risk as it returned a possibility value of 0.
2. The numeric level/s of transactional risk is dependent on the inputs and vary accordingly with a change of values. For example, let us consider the same TFAIC of Figure 7.7, but in this case let us consider the MIC is $5,000. Accordingly, the possibility distribution of transactional risk is shown in Figure 7.9.
3. From point 2, it can be seen that sometimes the convolution process of the inputs results in one of the focal element of transactional risk being greater than 100. In such scenarios, the resultant focal element is considered as 100,

as the UoD for that variable is defined till that range. If the convolution process results in having one of the focal elements of transactional risk as 100 in more than one instance, then, in each case, it is considered as 100 and its possibility value is determined from equation 7.6.

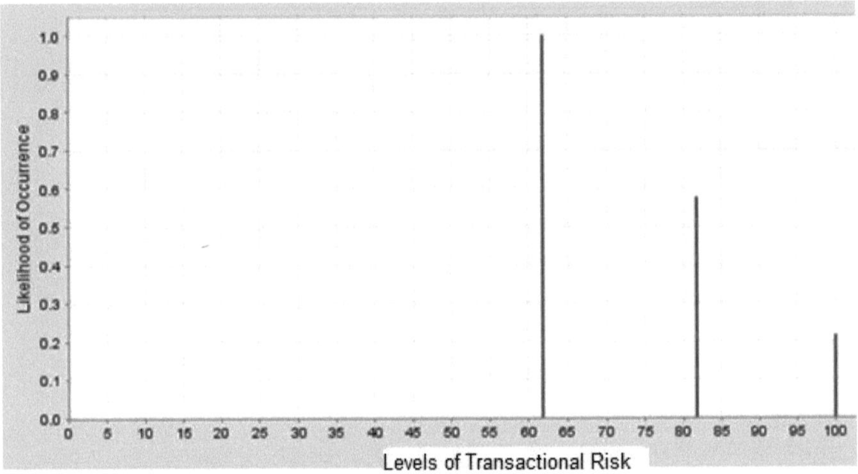

Fig 7.9 Possibility distribution of Transactional Risk with different inputs in agent 'A's business activity with agent 'B'

The above process gives the risk assessing agent the different numeric level/s of transactional risk possible in forming a business association with the risk assessed agent. In the next sections, we utilize fuzzy sets theory to ascertain linguistically the severity and likelihood of transactional risk in forming a business association.

7.10 Choice of Fuzzy Inference System to Ascertain the Linguistic Level/s of Transactional Risk in a Business Activity

As mentioned in [12], Fuzzy Logic assists in knowledge representation under conditions of uncertainty. It also assists in reasoning about the parameters by approximating them, rather than considering them as fixed values. This is achieved by fuzzyfying the inputs and output/s parameters according to the level of quantification of those parameters to the defined membership functions. The degree or level of quantification is represented by attaching probabilities to the fuzzy sets or predicates. The computation process of a fuzzy inference system is dependent on the type chosen to model it. There are two ways by which a fuzzy inference system can be modeled to the given problem:

(a) Mamdani Approach and
(b) Takagi and Sugeno Approach

Both these approaches assist in knowledge representation under conditions of uncertainty by approximating the analysis. Even though the workings of both these approaches vary, the series of steps in them can be divided into three broad categories. These are: a) input stage; b) computation stage; and c) output stage. We give a brief explanation of each step in each approach.

7.10.1 Input Stage

The input stage of both the above mentioned approaches is similar. In this stage, the inputs and their universe of discourse (UoD) are defined. This is followed by defining the fuzzy predicates and the membership functions. Furthermore, the input values are fuzzified according to the level of their quantification to those membership functions.

7.10.2 Computation Stage

The computation stage of both these approaches is responsible for knowledge representation under conditions of uncertainty. This stage varies according to the approach chosen for modeling the fuzzy inference system. Even though both of them utilize rules to process the approximated inputs and compute the output, the structure and how those rules are formed are different in each approach. In the Mamdani approach, the rules are formed in the form of the IF-THEN rules where the inputs and output (antecedent and precedent) are represented by the defined fuzzy predicates. Logical operators like AND, OR are used to join the predicates and compute the output fuzzy set. In the Takagi Sugeno approach, even though the rules are represented in the form of IF-THEN statements, the consequent part of the rule is not represented by a fuzzy predicate but a linear function. The order of the linear function in the consequent part of the rule depends on the type of problem that is being modeled and solved.

7.10.3 Output Stage

The output stage is the conclusion or the knowledge of the fuzzy inference system by processing the inputs. Both of these approaches vary in the output stage and represent it in different ways. In the Mamdani approach, the output is represented in the form of fuzzy predicate/s on the defined membership function and universe of discourse; whereas in the Takagi-Sugeno approach, the output is represented as a linear crisp value.

So, the type of output required according to the problem objective will lead to choosing the type of approach to model the fuzzy inference system. As discussed earlier, our objective to utilize the fuzzy inference system is to determine the semantic representation of transactional risk in forming a business activity. Depending upon the interaction specific characteristics, there might be more than one level of severity of transactional risk, and all these level/s have to be identified in order to make an informed representation relating to its presence in the business activity. Furthermore, the likelihood of the occurrence of these levels has to be determined on the defined fuzzy predicates to have a semantic representation of transactional risk. Such type of output is best achieved by using the Mamdani approach. Another difference in the working of the Mamdani and Takagi-Sugeno approach is that the Takagi-Sugeno approach requires the system to be trained first in order for it to behave according to the required environment constraints and conditions for achieving an output. This approach works well when there is sufficient data available that captures all the possible scenarios, where part of the data can be divided to initially train the model and the other part can be utilized to test and validate it. However, in situations where there is insufficient data to train the system or if the system encounters inputs during the computation stage with which it was not familiar in the learning phase, then the given output may include a bias or an error. This error may propagate in the further stages (for example risk evaluation and risk management phases) which must be avoided. So, to avoid these issues, we utilize the Mamdani approach in our framework to model the fuzzy inference system. In the next section, we explain the detailed steps in determining the linguistic level/s of transactional risk in the business activity by using the Mamdani approach.

7.11 Steps to Ascertain the Linguistic Level/s of Transactional Risk in a Business Activity

The series of steps to be carried out for developing a fuzzy inference system that ascertains the linguistic level/s of transactional risk are as follows:

7.11.1 Define the Inputs, Output/s and Their Universe of Discourse

The aim of this step is to:

(a) identify the input and output variables to ascertain the linguistic level/s of transactional risk, and
(b) define their universe of discourse.

As discussed in Section 7.7, while determining the numeric level/s of transactional risk in a business activity, two inputs, namely Loss of Investment Probability (LOIP) and Possible Consequences of Failure (PCF) are considered. These same inputs are considered while determining the linguistic level/s of transactional risk. Furthermore, in Section 7.7, we defined the universe of discourse of the input variables to ascertain the numeric level of transactional risk. We utilize the same range as the UoD of those variables while developing the fuzzy inference system to ascertain the linguistic level/s of transactional risk.

7.11.2 Define the Fuzzy Sets or Predicates for the Input and Output Variables

The values of both the input variables (namely LOIP and PCF) are crisp numerical values, which have to be fuzzified into linguistic terms. So, in order to transform each variable from crisp data to fuzzy variables, fuzzy sets or predicates firstly have to be defined. A fuzzy set or predicate is a linguistic phrase that is used as a semantic label to represent a range of that variable, which best matches its description. The predicates for a variable should be defined such that they cover the whole numerical range, or the scope or the universe of discourse within which the variable extends. In our problem, in order to have an informed categorization and representation of the input/s and output, we consider that each of the variables (inputs and output) is divided into 6 fuzzy sets or predicates, as shown in Figure 7.10.

Predicates or the Fuzzy sets	Acronym
Extremely Low	'EL'
Low	'L'
Low Medium	'LM'
Medium High	'MH'
High	'H'
Extremely High	'EH'

Fig. 7.10 Fuzzy sets or predicates for the inputs and output variable

7.11.3 Define the Membership Functions of the Fuzzy Sets

Once the different predicates for a variable have been defined, the next step is to determine their range or boundaries on the universe of discourse. While defining the membership functions, it is important to consider the concept of partial truth and overlapping values across the different fuzzy sets. The membership function of a variable is used to transform each crisp input value into a fuzzy variable, by utilizing the predicates and determining the degree of membership of the input value to those predicates. There are various shapes by which the membership function of a variable can be plotted, for example g-bell curve, trapezoidal function, triangular function, Gaussian etc. Each of these curves is shown in Figures 7.11 – 7.13, respectively.

Depending on which type of shape is chosen, the membership function of all the fuzzy sets in the input and output variables should be plotted. An example of the membership functions of each input and output variable used in our approach is shown in Figures 7.14 -7.15 and Figure 7.16 respectively [10].

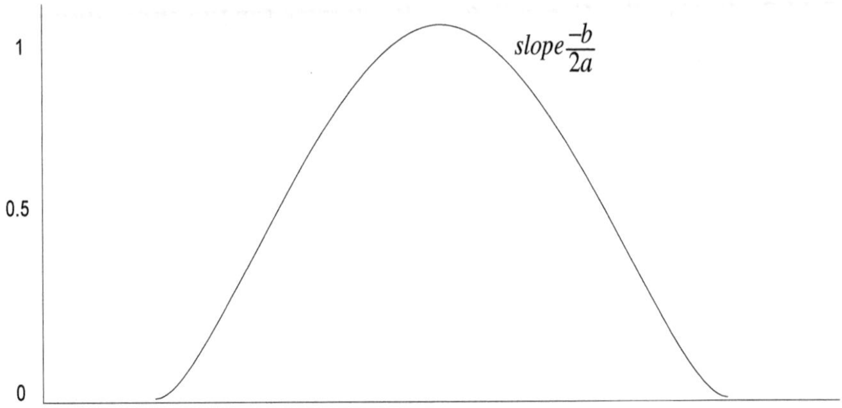

Fig. 7.11 Bell shaped membership function

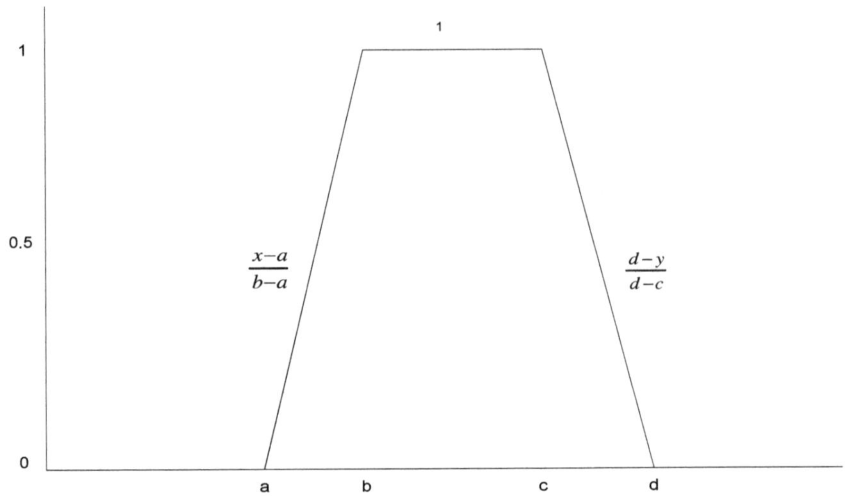

Fig. 7.12 Trapezoidal membership function

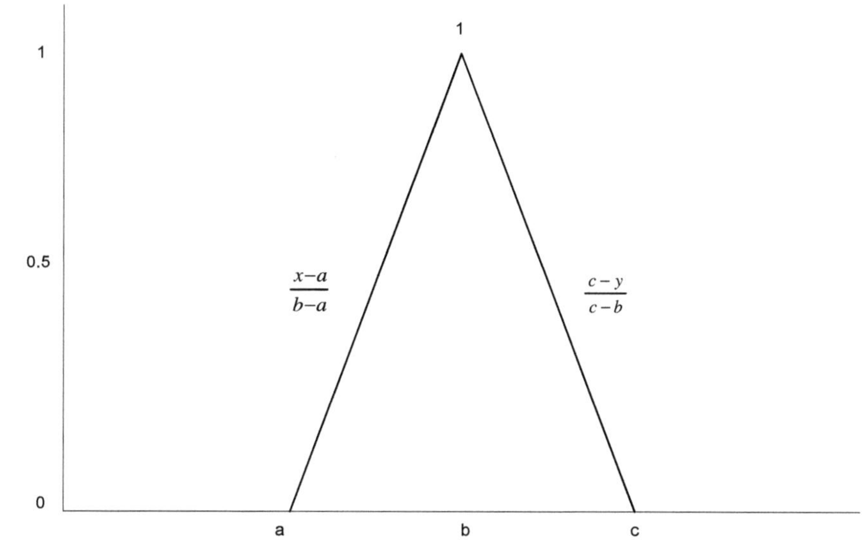

Fig. 7.13 Triangular membership function

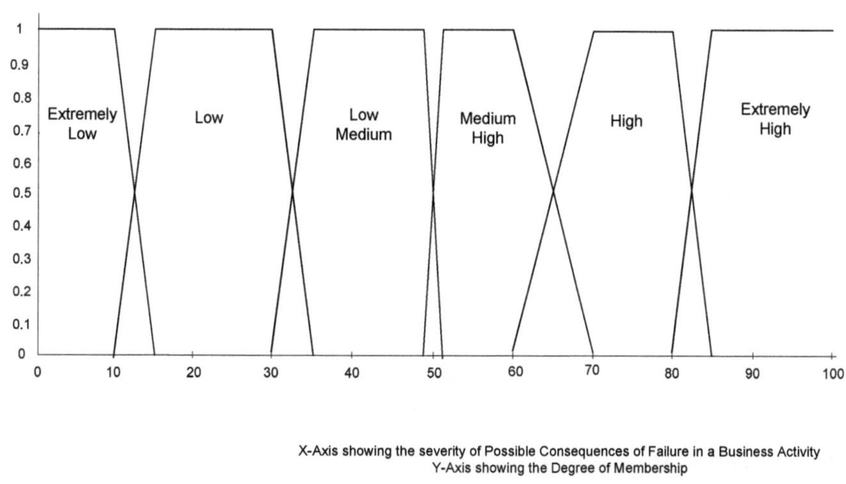

Fig. 7.14 Membership function of the Input - Possible Consequences of Failure (PCF)

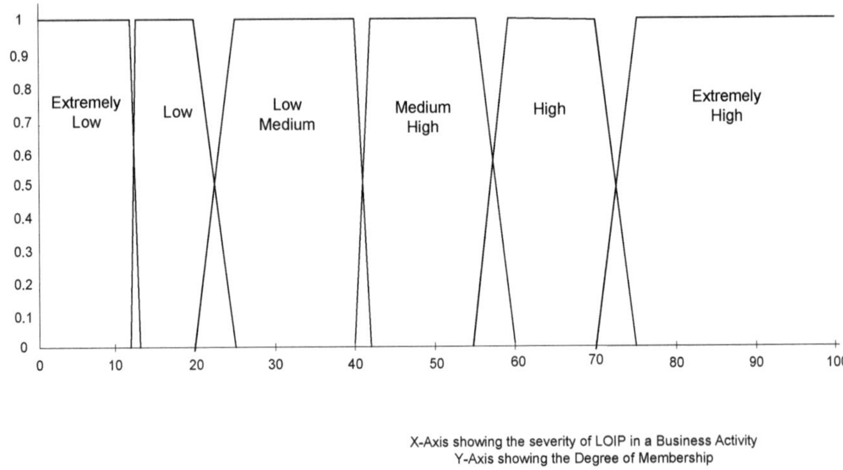

Fig. 7.15 Membership function of the Input - Loss of Investment Probability (LOIP)

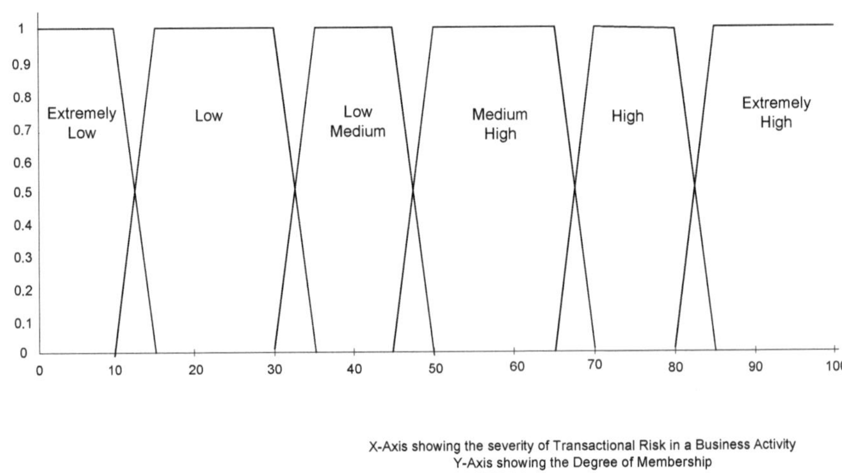

Fig. 7.16 Membership Function of the Output - Transactional Risk (TR)

7.11.4 Fuzzify the Inputs by Determining Their Degree of Membership to the Membership Function

The degree of membership (DOM) of a variable to a fuzzy set or predicate defines the magnitude of participation of that variable within that fuzzy set. The degree of membership of a variable is determined by plugging the selected input parameter into the horizontal axis of its UoD and projecting it vertically at the point of intersection with the fuzzy set/s. The point on the y-axis to which the input variable intersects a fuzzy set represents its degree of influence or the DOM to that fuzzy set. By using the membership function, the focal elements of each input variable (determined in Section 7.8) should be transformed into fuzzy sets, based on the level of truth to which the input variable quantifies the predicates.

To transform a focal element 'x' of an input linguistic variable to the defined fuzzy sets, the possibility to which that element 'x' corresponds with the predicates of that input variable are determined. This is done by considering the overlap between the degree of evidence of the focal element 'x', with the degree of membership to which that input value 'x' corresponds to a particular predicate from the membership function for that input variable. Mathematically, the possibility that the fuzzy set or predicate 'A' of an input linguistic variable will occur, based on the degree of evidence of input 'x', is given by [4, 9]:

$$\Pi(A) = \max\{\min[\pi(x), DOM_A(x)]\} \tag{7.7}$$

Equation 7.7 is repeated for each focal element '*x*' of an input variable to determine the possibility of occurrence of a fuzzy set or predicate 'A'. Once all the input variables have been transformed to their corresponding fuzzy sets, the next step is for them to be processed in the inference engine to draw a conclusion on the UoD and membership of the output linguistic variable.

7.11.5 Define the Inference System of the Fuzzy Model

After determining the possibility of the quantification of each input variable to the fuzzy sets, they must be fed to the inference engine to draw a conclusion on the output membership function. The inference engine consists of fuzzy rules by which conclusions are drawn. The rules which we use in our model are of the IF-THEN structure. We abbreviate the part of the rule between IF and THEN as the 'LHS' (Left-hand side) of the rule. The LHS of the rules consist of the fuzzy sets of the inputs, determined based on their focal elements. Similarly, the implication part of the rule, which is associated after the THEN operator, is abbreviated as the 'RHS' (Right-hand side) of the rule. The implication part of the rule gives the level of severity of transactional risk in the business activity, based on the computations performed by the inference engine. The LHS of the rules is combined by a logic operator. We use the AND operator as we want to consider the intersection between the input fuzzy sets, while implicating the likelihood of transactional risk by utilizing those sets. In our fuzzy inference model, there are two inputs and each input is further defined by six predicates. Hence, the total number of homogenous rules in our system is 36 as shown in Figure 7.17 [10].

	PCF		LOIP		TR
If	EL	and	EL	then	L
If	L	and	EL	then	L
If	LM	and	EL	then	LM
If	MH	and	EL	then	MH
If	H	and	EL	then	H
If	EH	and	EL	then	EH
If	EL	and	L	then	L
If	L	and	L	then	LM
If	LM	and	L	then	MH
If	MH	and	L	then	H
If	H	and	L	then	EH
If	EH	and	L	then	EH
If	EL	and	LM	then	LM
If	L	and	LM	then	MH
If	LM	and	LM	then	H
If	MH	and	LM	then	EH
If	H	and	LM	then	EH
If	EH	and	LM	then	EH
If	EL	and	MH	then	MH
If	L	and	MH	then	H
If	LM	and	MH	then	H
If	MH	and	MH	then	EH
If	H	and	MH	then	EH
If	EH	and	MH	then	EH
If	EL	and	H	then	H
If	L	and	H	then	EH
If	LM	and	H	then	EH
If	MH	and	H	then	EH
If	H	and	H	then	EH
If	EH	and	H	then	EH
If	EL	and	EH	then	EH
If	L	and	EH	then	EH
If	LM	and	EH	then	EH
If	MH	and	EH	then	EH
If	H	and	EH	then	EH
If	EH	and	EH	then	EH

Fig. 7.17 Rules of the fuzzy inference system

7.11.6 Determine the Output of the Fuzzy Inference System

The output of the fuzzy inference system will be determined by the degree or strength to which each rule fires. The rules use the DOM of the fuzzy set of each input as the weighting factor to determine their influence on the output fuzzy sets.

The output of each rule must be aggregated to determine the different fuzzy sets that show the linguistic level/s of transactional risk in the business activity. The aggregation operator combines the output of the various rules corresponding to a fuzzy set into one, and determines the possibility of occurrence of that fuzzy set. There are different available operators to achieve this. In our approach, we use the multiple or additive aggregation operator. The multiple additive method determines the strength to which the output fuzzy set quantifies, by considering all the rules which produce the same fuzzy set and then utilizes a fuzzy additive principal bounded to a sum of 1.0. This is in contrast to the other approaches of aggregation, for example, the max-min method that considers all the rules which produce the same fuzzy set and then selects the maximum value from the rules as the strength to which that output fuzzy set quantifies. This ignores the output from the other rules that have a firing strength less than the maximum value.

If a scalar output of transactional risk has to be determined, then the fuzzy sets from the aggregation process must be 'defuzzified'. There are different approaches proposed for defuzzification. The most commonly used is the center of gravity or centroid method for defuzzification. In the centroid method, the scalar or crisp value of the output variable is computed by finding the center of gravity of the fuzzy set/s computed in the aggregation part by using Equation 7.8:

$$x' = \frac{\int \mu(x) x dx}{\int \mu(x) dx} \tag{7.8}$$

where:

$\mu(x)$ is the output fuzzy set after the aggregation of the individual implication results.

If the aggregated value from the rules spreads over more than one fuzzy set in the output membership function, then each of those membership functions must be considered while computing the centroid. In our approach, as our aim is to capture the different level/s of severity in the occurrence of transactional risk in a business activity, we do not defuzzify the output fuzzy sets. Rather, we ascertain the possibility of occurrence of each fuzzy set along with the degree of membership of that set.

7.12 Ascertaining the Linguistic Level/s of Transactional Risk in a Business Activity

Let us consider the previously discussed example of a business activity between agent 'A' and agent 'B' discussed in Section 7.9. Let us consider that the MIC of the risk assessing agent 'A' is $5,000 on the TFAIC of Figure 7.7. Accordingly, by using the fuzzy sets, membership functions and rule base defined in the previous section, the linguistic level/s of transactional risk in forming a business activity is shown in Figure 7.18.

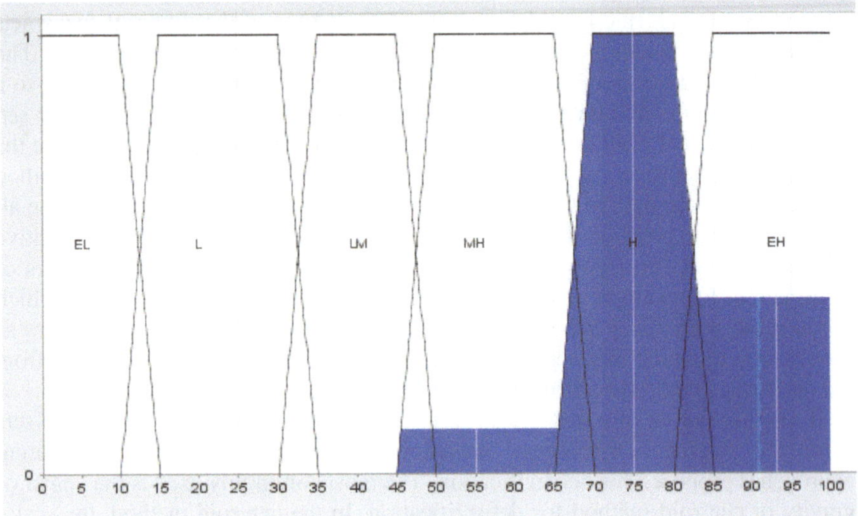

Fig. 7.18 Linguistic level/s of transactional risk in a business activity

Figure 7.18 shows the different linguistic fuzzy sets along with their DOM in forming a business association. In contrast with the numeric level/s of transactional risk, the linguistic level/s represents the fuzzy sets along with their DOM that shows in semantic terms, the different level/s of transactional risk in the business activity. This will assist the risk assessing agent in having a semantic representation of the transactional risk in the business activity. The next step after determining the level/s of transactional risk is the risk evaluation phase. In this phase, the risk assessing agent has to make a decision about whether or not the determined level of transactional risk is 'acceptable' to it. We discuss that process further in the next chapter.

7.13 Recapitulation

In this chapter, we proposed a methodology by which the risk assessing agent can combine the different input constituents that it had determined for risk analysis and ascertain the transactional risk in forming a business activity with a risk assessed agent. The proposed approach determines the transactional risk numerically and linguistically. The numeric representation utilizes possibility theory in determining the different level/s of severity of transactional risk and the possibility of occurrence of those levels. The linguistic representation utilizes the fuzzy inference system to determine, in semantic terms, the level/s of severity of transactional risk and their degree of occurrence. The next phase in the process of risk analysis is evaluating the determined levels of transactional risk. We discuss this further in the next chapter where we propose a methodology that assists the risk assessing agent to evaluate the determined levels of transactional risk present in the business association with the risk assessed agent.

References

1. Meadow, C.: Text Information Retrieval Systems. Academic Press, San Diego (1992)
2. Embley, D.W.: Towards Semantic Understanding - An Approach Based on Information Extraction Ontologies. In: 15th Australasian Database Conference (ADC 2004), Dunedin, New Zealand, pp. 3–12 (2004)
3. Weiss, N.A., Holmes, P.T., Hardy, M.: A Course in Probability. Addison-Wesley (2005)
4. Darby, J.L.: Estimating Terrorist Risk with Possibility Theory. Los Alamos National Lab, US Department of Energy, (US) Manuscript number LA-12179 (2004)
5. Zadeh, L.A.: Fuzzy Sets As a Basis For A Theory of Possibility. Fuzzy Sets and Systems 100, 9–34 (1999)
6. Dubious, D., Prade, H.: Possibility Theory, Probability Theory and Multiple-Valued Logics: A Clarification. Annals of Mathematics and Artificial Intelligence 32, 35–66 (2001)
7. Dubois, D.: Possibility theory and statistical reasoning. Computational Statistics & Data Analysis 51, 47–69 (2006)
8. Zimmerman, H.-J.: Fuzzy Set Theory and Its Applications. Kluwer Academic Publishers (1991)
9. Berkan, R.C., Trubatch, S.L.: Fuzzy Systems Design Priciples. IEEE Press, New York (1997)
10. Hussain, O.K., Dillon, T.S., Hussain, F.K., Chang, E.: Transactional Risk-based Decision Making System in E-Business Computing. Computer Systems Science and Engineering 25, 15–25 (2010)
11. Dubois, D., Prade, H., Farreny, H., Martin-Clouaire, R., Testemale, C., Harding, E.F.: Possibility Theory: An Approach to Computerized Processing of Uncertainty. Plenum Press, New York (1988)
12. Yager, R.R., Zadeh, L.A. (eds.): An Introduction To Fuzzy Logic Applications In Intelligent Systems. Kluwer Academic Publishers (1992)

Chapter 8
Risk Evaluation Phase: Transactional Risk Evaluation and Decision Making in Business Activities

8.1 Introduction

In the previous chapter, we proposed an approach by which the risk assessing agent ascertains the different representations and the different severity level/s in the occurrence of perceived transactional risk (transactional risk) in a business activity. The next step is for the risk assessing agent to evaluate the determined level/s of transactional risk to make a decision about his future course in forming a business association with a risk assessed agent. Alternatively, the risk assessing agent may have to make a decision about choosing an agent from a set of risk assessed agents with whom to form a business association. This occurs in the risk evaluation phase, which is the next stage in the process of risk analysis. In the risk evaluation phase, the risk assessing agent utilizes the determined severity level/s of transactional risk in forming a business association with the risk assessed agent; and then evaluates this to determine whether or not it is within *acceptable* or *tolerable* limits. In other words, by evaluating the determined level/s of transactional risk, the risk evaluation phase assists the risk assessing agent to make an informed interaction-based decision about the risk assessed agent.

In this chapter, we present an approach by which the risk assessing agent can evaluate the determined level/s of transactional risk in forming a business activity and then utilize it to make an informed interaction-based decision. However, before transactional risk can be evaluated, it is important to identify and capture the different factors which introduce variability in it. We discuss this further in the next section.

8.2 Variability to Be Identified and Captured While Evaluating Transactional Risk

When evaluating the different level/s of transactional risk in the business activity, it is important to identify and measure several variable factors when making an assessment about a course of action with a risk assessed agent. To illustrate with

O.K. Hussain et al.: Risk Assessment & Management in the Networked Economy, SCI 412, pp. 217–241.
springerlink.com

an example, let us consider agent 'A's interaction with agent 'B' in forming a business association, as shown in Figure 7.3. As discussed in the previous chapter, on a scale of 0-100, it is a commonly understood notion that the closer the value towards 100, the higher the severity of the factor being represented. In order to correctly understand the severity of the level/s of transactional risk, semantics are assigned to the range of values over which transactional risk is determined. Even though this assists in understanding the determined level/s of transactional risk, it does not assist in measuring these variations in the assessment of the determined level/s of transactional risk. These variations are dependent on the interpretation of the severity level/s and magnitude of occurrence of transactional risk and the particular time in which the business activity is being considered. This is discussed further in the next sub-sections.

8.2.1 Variation in the Interpretation of Transactional Risk according to the Level/s of Severity and Likelihood of Its Occurrence

An important factor to be considered during transactional risk evaluation is to understand how the risk assessing agent interprets the determined severity level/s of transactional risk in the business activity. For example, as shown in Figure 8.1, Agent 'A' may consider the level of severity of transactional risk (say 20% and

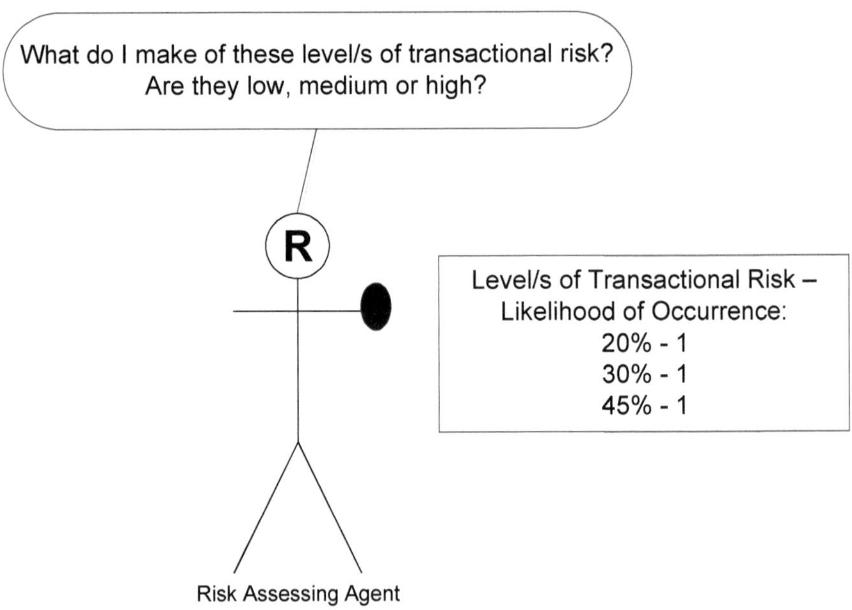

Fig. 8.1 Variation in the interpretation of transactional risk in the business activity

30%) as low or acceptable but the severity level of transactional risk (say 40%) as *high* in the business activity. Or, if the likelihood of occurrence of the severity level of transactional risk (40%) is 0.2, then Agent 'A' may consider this to be acceptable, compared to when it is 1. Conversely, another risk assessing agent, (say agent 'C') may consider the same severity level of risk which agent 'A' considers *high* to be *moderate,* as shown in Figure 8.2.

Such differences in the evaluation of the severity of risk will lead to different outcomes in the risk evaluation and risk management phases; hence they need to be accurately measured in order to have a correct understanding of the risk evaluation.

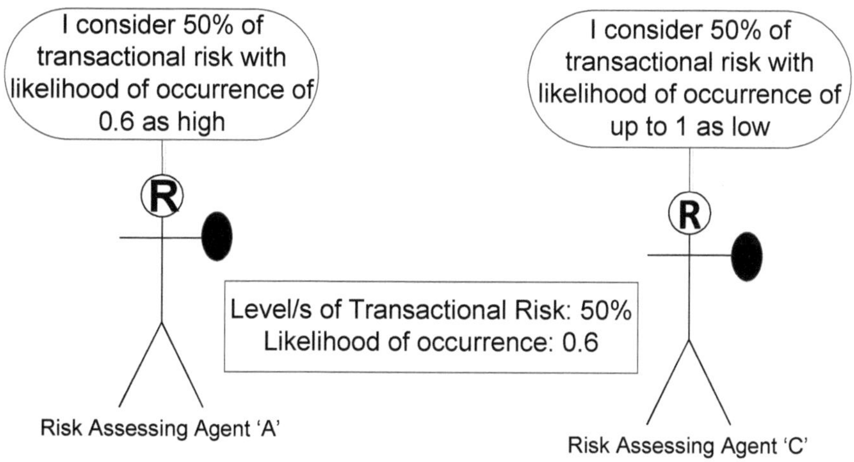

Fig. 8.2 Variation in the interpretation of transactional risk severity level/s between different agents

8.2.2 Variation in the Interpretation of Transactional Risk according to the Time of Consideration

Another factor that introduces variation in the interpretation of transactional risk in the business activity is time. To illustrate with an example, let us consider the interaction scenario with risk assessing agent 'C', as discussed in the previous section and that after evaluating the severity level/s of transactional risk, agent 'C' initiates a business activity with risk assessed agent 'B' in January 2008 for a period of two years. From January 2008 till June 2008, agent 'C' considers the severity level/s of transactional risk of 50% with the likelihood of occurrence of 1, as low. However in July 2008, due to the global financial crisis and the resulting circumstances which affected agent 'C', his outlook towards the transactional risk changed and he now considers the same severity level of transactional risk to be 'high', as illustrated in Figure 8.3. As a result of this, the determined level/s of risk in the business activity needs to be re-evaluated and appropriate steps need to be taken to ensure the successful completion of the business activity.

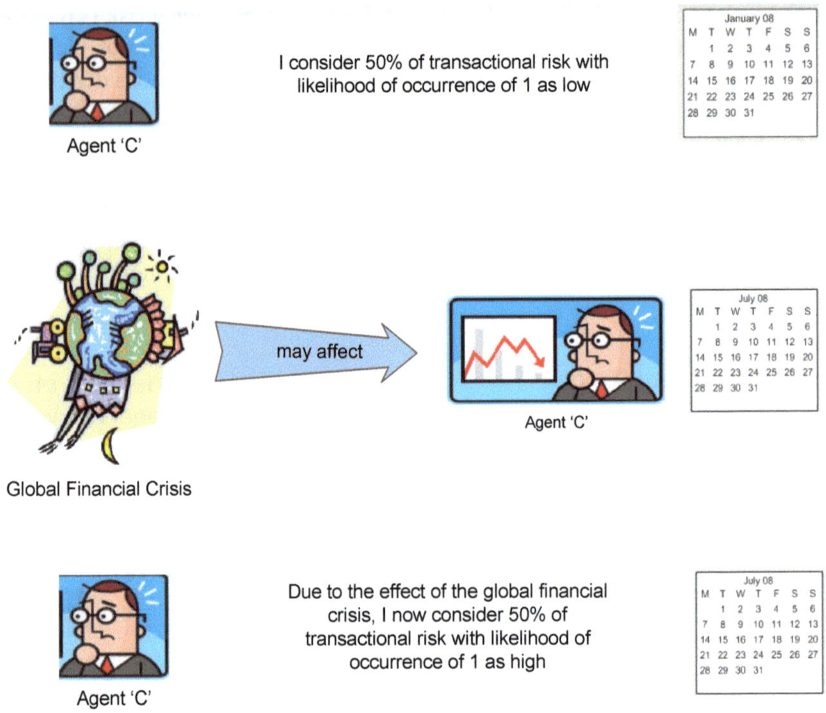

Fig. 8.3 Variation in the interpretation of transactional risk level/s according to time

It is important to consider both these factors in the process of transactional risk evaluation and hence, these need to be identified appropriately. This is done by measuring the risk propensity or risk attitude of the risk assessing agent in the business activity. We discuss that further in the next section.

8.3 Importance of Risk Propensity While Undertaking Transactional Risk Evaluation

Risk propensity defines the risk assessing agent's risk-taking nature or his risk attitude in the business activity at a certain period of time. In other words, it also defines the risk assessing agent's current tendency towards taking risks in an activity and determines how he will behave or react to the severity level of transactional risk, and based on this, which levels of transactional risk in the business activity are acceptable to him and which are not. So, in a business activity, the risk propensity of the risk assessing agent is the main factor in determining what his interaction-based decision output is going to be [1]. The output of this phase will form the input for the next phase of risk management.

As discussed in the previous section, it is possible that no two risk assessing agents will have the same risk propensity or risk attitude at a certain point in time, and with this variation in their risk attitudes, the output of the risk evaluation and their approach to decision-making in the business activity varies. Also, as previously discussed, the risk attitude of a risk assessing agent might not remain the same throughout the business activity; rather, it may vary. Hence whilst transactional risk evaluation is important, it is also very important for the risk assessing agent to first ascertain accurately his risk propensity at that given period of time and then to determine the impact this has on the level/s of transactional risk determined in the business activity. This will assist the risk assessing agent make an informed decision regarding his future course of interaction in a business activity. In the next section, we discuss the different level/s of risk propensity which capture the risk attitude of the risk assessing agent.

8.4 Different Levels to Measure the Risk Propensity of a Risk Assessing Agent

The risk propensity of the risk assessing agent should be determined from a spectrum or range which specifies all the different risk-taking natures or risk attitudes possible. We define the spectrum of the risk propensity of the risk assessing agent in three broad categories, as explained below.

8.4.1 Risk Averse

Risk Averse is defined as that attitude of the risk assessing agent where he wants to take only minimal risk in the business activity. In other words, agents with such risk propensity proceed in the business activity only if the severity level of transactional risk in the business activity is either nil or minimal.

8.4.2 Risk Neutral

Risk Neutral is defined as that attitude of the risk assessing agent where he wants to proceed in the business activity only if the advantages it will achieve will outweigh the costs involved. In other words, agents with a risk propensity of risk neutral do not totally avoid risk as done by those agents with a risk averse nature, and will accept transactional risk in the interaction to a certain severity level.

8.4.3 Risk Taking

Risk Taking is defined as that attitude of the risk assessing agent where he is prepared to take high levels of transactional risk and is ready to interact with a risk assessing agent, even when this high severity level of transactional risk is present in the business activity. In other words, agents with a risk propensity level of risk taking are ready to accept different severities of the level/s of transactional risk in the business activity.

The risk attitudes of the risk assessing agent in terms of accepting the severity level/s of transactional risk in the business activity can be arranged in the ascending order of RA < RN < RT. To evaluate the severity level/s of transactional risk in a business activity, the risk assessing agent has to first choose his risk propensity level or risk attitude from the spectrum and then consider its impact on the determined level/s of transactional risk. However, before this can be done, the boundaries or ranges of the risk propensity level/s of the risk assessing agent need to be defined. Furthermore, it is possible that the risk propensity level of the risk assessing agent might not correspond crisply to a certain level but there might overlap across two different adjacent levels, for example, risk averse and risk neutral or risk neutral and risk taking. To take such scenarios into account, we propose a fuzzy inference system that initially captures the fuzziness of the risk attitude of the risk assessing agent, and then utilizes it to accurately determine its impact on the severity level/s of transactional risk in the business activity. In other words, the proposed fuzzy inference system evaluates the effect of the determined severity level/s of transactional risk according to the risk propensity of the risk assessing agent, and semantically recommends the risk-based decision as output, prior to forming an interaction with that agent. Furthermore, using a fuzzy inference system allows the output of the transactional risk evaluation phase to be expressed semantically, enabling a more accurate analysis to be conducted which assists decision-making.

In the following sections, we define and develop a fuzzy inference model which quantifies the risk propensity of the risk assessing agent and then determines its impact on the severity level/s of transactional risk in interacting with the risk assessed agent, in order to recommend an interaction-based decision to the risk assessing agent.

8.5 Output of the Transactional Risk Evaluation Phase

As mentioned earlier, the transactional risk evaluation phase evaluates the level/s of transactional risk in the business activity, according to the risk propensity of the risk assessing agent. However, as mentioned in transactional risk assessment in Section 7.4, the output of transactional risk evaluation phase too, should be expressed in a way that is best understood by the risk assessing agent to enable a clearer and accurate overview of risk in the risk management phase. Furthermore, it is important to note that the process of transactional risk evaluation is not a one-off process; rather it can be an iterative process, as shown in Figure 3.19. So, it is important to determine how the output of the transactional risk evaluation phase will be expressed, so that it can be understood in all stages.

There are various ways to express the output of the transactional risk evaluation phase. However, in our approach, as we use a fuzzy inference system to ascertain the impact of the risk propensity of the risk assessing agent on the determined level/s of transactional risk, we express the output of the transactional risk evaluation phase on defined fuzzy predicates. There are many examples of fuzzy predicates that can be utilized for this, for example *accept*, *reject, acceptable, unacceptable* etc. However, we want to suitably express the output of the transactional risk evaluation phase in such a way that it assists the risk assessing agent to make an informed interaction-based decision in relation to the business activity. As discussed in Section 1.6, an informed interaction-based decision in the context of the discussion of this book is defined as a proactive decision that will lead the risk assessing agent in the achievement of the goals for which the collaboration in the business endeavor was formed. So, the fuzzy predicates that we choose to represent the output of the transactional risk evaluation phase are as follows:

8.5.1 Proceed

The fuzzy set *Proceed* indicates to the risk assessing agent that the determined severity level/s of transactional risk in the business activity along with their magnitude of occurrence is acceptable, according to the risk propensity of the risk assessing agent. Based on the analysis, the risk assessing agent can proceed in forming a business activity with the risk assessed agent.

8.5.2 Don't Proceed

The fuzzy set *Don't Proceed* indicates to the risk assessing agent that the determined severity level/s of transactional risk in the business activity along with their magnitude of occurrence is unacceptable, according to the risk propensity of the risk assessing agent. Based on the analysis, the risk assessing agent should not proceed in forming a business activity with the risk assessed agent and should iterate through the phases of risk management and risk evaluation again.

These two fuzzy predicates represent the two possibilities for the risk assessing agent to consider while making an informed interaction-based decision. Depending on the degree of quantification of these fuzzy sets, the risk assessing agent will have a better understanding of the severity of the risk and the steps to be taken in the next phase of transactional risk management.

8.6 Fuzzy Inference Model for Transactional Risk Evaluation in the Business Activity

An overview of the fuzzy inference system, according to the inputs which it receives and the output that it gives to evaluate the transactional risk and assist the risk assessing agent to make a decision on a business activity with the risk assessed agent, is shown in Figure 8.4.

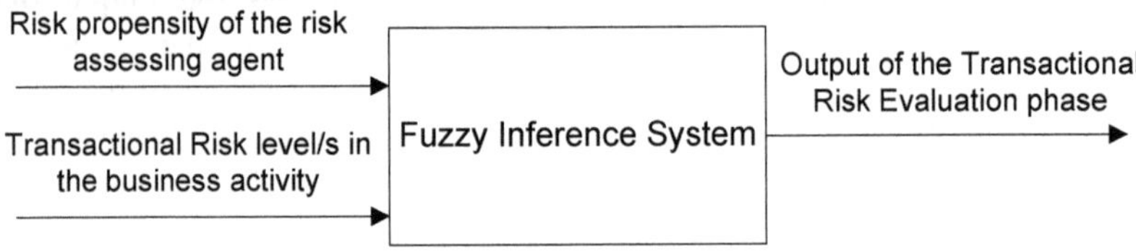

Fig. 8.4 Overview of the fuzzy inference system for Transactional Risk Evaluation

8.6.1 Representation of Transactional Risk Used in the Evaluation

In the previous chapter, we discussed the two representations of transactional risk, namely numeric and linguistic, used to measure its severity and likelihood of occurrence in the business activity. During transactional risk evaluation, we propose that the linguistic representation of transactional risk is used when it is input into the fuzzy inference system to evaluate it. This is because the linguistic representation of transactional risk has semantics associated with the occurrence of the different severity level/s and their DOM, which can be utilized better by the fuzzy inference system to evaluate them and ascertain the semantic decision output in the business activity.

The operational structure of a fuzzy inference model is discussed in the last chapter. In the next section, we define the universe of discourse and membership functions of the input and output variables to evaluate the severity level/s of transactional risk in the business activity.

8.7 Universe of Discourse and Membership Functions of the Variables for Transactional Risk Evaluation

In this section, we define the universe of discourse and the membership functions of the input and output variables used for transactional risk evaluation.

8.7.1 Fuzzy Sets and the Membership Function for the Input: Risk Propensity (RP) of the Risk Assessing Agent

The input to the system *Risk Propensity (RP)* of the risk assessing agent represents the quantified risk attitude or the risk-taking nature of the risk assessing agent in the business activity. In Section 8.4, the spectrum over which the possible risk attitude of the risk assessing agent extends is defined. In order to consider the levels from that spectrum in our fuzzy inference model, we define the universe of discourse (UoD) of *Risk Propensity* of the risk assessing agent in the range of 1-5; {1, 2, 3, 4, 5} where each element represents a numeric value and is unit-less.

To classify different fuzzy sets for the input variable *Risk Propensity* of the risk assessing agent, we divide the universe of discourse into 3 predicates, *Risk Averse, Risk Neutral and Risk Taking*. The defined membership functions for the linguistic input *Risk Propensity* of the risk assessing agent are as shown in Figure 8.5. The membership function is used to determine the strength to which a value from the UoD of the input variable quantifies to the defined predicates of the risk attitude of the risk assessing agent.

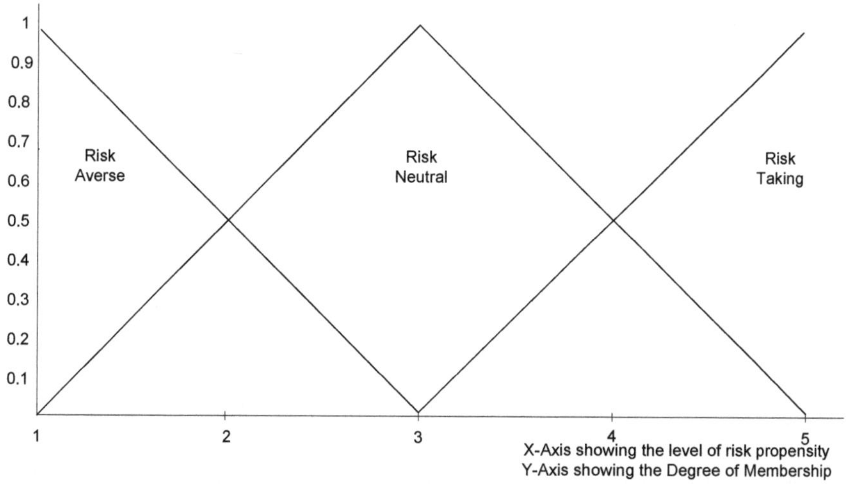

Fig. 8.5 Membership function for the input Risk Propensity

8.7.2 *Fuzzy Sets and the Membership Function for the Input: Transactional Risk (TR) in the Business Activity*

The input variable *Transactional Risk (TR)* in the interaction represents the determined severity level/s and magnitude/s of occurrence of transactional risk in the business activity. As mentioned earlier, to determine semantically the output of the risk evaluation phase and the recommended decision, the risk assessing agent utilizes the linguistic representation of transactional risk. Thus, we consider the universe of discourse, predicates and membership function for the variable *Transactional Risk* as defined in Figure 7.14.

8.7.3 *Fuzzy Sets and the Membership Function for the Output: Transactional Risk Evaluation - Recommended Risk-based Decision (RRD)*

The fuzzy inference system, based on the inputs given to it, evaluates the transactional risk and computes an output specifying the *Recommended Risk-based Decision (RRD)*. We consider a range of 0-10; {0, 1, 2, ……, 10} as the universe of discourse (UoD) of this variable. The membership function for the output *Recommended Risk-based Decision* in the business activity is defined as shown in Figure 8.6 [2] spread over the universe of discourse for the fuzzy variable.

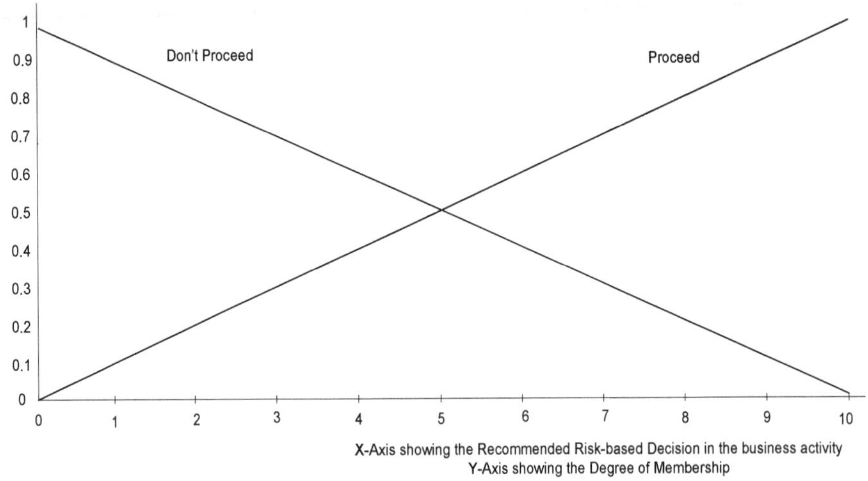

Fig. 8.6 Membership function for the output - Recommended Risk-based Decision

Once the input variables have been transformed into fuzzy sets by using their defined membership function, they must then be processed in the inference engine to determine the impact of the risk propensity of the risk assessing agent on the determined severity level/s of transactional risk. However, as mentioned earlier, before this can be done, the boundary or range of the maximum acceptable level/s of transactional risk to each risk propensity level should be defined, as it is against these boundaries that the transactional risk in a business activity will be evaluated, according to the risk propensity of the risk assessing agent. We discuss this further in the next few sections.

8.8 Definition of Key Terms Used in Transactional Risk Evaluation

In this section, we define the key terms which we use in this stage in the process of transactional risk evaluation.

8.8.1 Maximum Acceptable Transactional Risk Level (MARL)

Maximum Acceptable Transactional Risk Level represents that fuzzy set or predicate from the transactional risk level/s which are acceptable to the risk assessing agent, according to his risk propensity level.

8.8.2 Maximum Risk Propensity of the Risk Assessing Agent (MRP)

The Maximum Risk Propensity of the Risk Assessing Agent (MRP) defines the risk attitude or risk propensity of the risk assessing agent.

8.8.3 *Current Risk Level (CRL)*

Current Risk Level (CRL) is the fuzzy predicate of transactional risk which is currently being considered by the risk evaluation fuzzy inference system.

8.8.4 *Current Risk Attitude (CRA)*

Current Risk Attitude (CRA) is the fuzzy predicate of the risk propensity of the risk assessing agent which is currently being considered by the risk evaluation fuzzy inference system.

8.8.5 *Possible to Proceed in the Interaction at This Stage (Poss)*

Possible to Proceed in the Interaction at this Stage (Poss) is an intermediate variable that represents whether or not a certain fuzzy set of the transactional risk (CRL), along with its DOM, is acceptable to the risk assessing agent, according to his risk propensity level (CRA).

In the next section, we define the MARL of the risk assessing agent, according to the different risk propensity levels.

8.9 MARL in the Case of Crisp Risk Propensity Level

To ascertain the MARL of the risk assessing agent, two important prerequisites that need to be achieved are as follows:

(a) accurately determining the risk propensity of the risk assessing agent: as discussed earlier, with changes in the risk propensity levels, the way the risk assessing agent assesses transactional risk changes. To clarify this with an example, a risk assessing agent with a risk propensity nature of Risk Averse (RA) = 1, will analyze transactional risk in interacting with a risk assessed agent to be acceptable or not in a different way compared to the same risk assessing agent assessing the same transactional risk level/s if his risk propensity nature is a combination of Risk Averse (RA) and Risk Neutral (RN) levels. So, the accurate risk propensity level of a risk assessing agent has to be measured before ascertaining the impact of transactional risk.

(b) determining the maximum acceptable level of transactional risk by the risk assessing agent: during transactional risk evaluation, the linguistic representation of transactional risk has to be evaluated against the risk attitude of the risk assessing agent, according to the level/s to which it corresponds in relation to the risk propensity spectrum. To do this, it is important to define the level/s of transactional risk which a risk assessing agent will accept, according to a certain risk propensity level. For example, with a risk propensity level of Risk Averse, what is the maximum acceptable level of transactional risk (MARL) that the risk assessing agent will accept, as it is against this level that the risk evaluation process will be carried out.

In Section 8.4, we defined the risk propensity spectrum of the risk assessing agent which has three broad categories, namely risk averse, risk neutral and risk taking. In Figure 7.14, we defined the different fuzzy sets used in the linguistic representation of transactional risk. Arranged in ascending order of severity, they are EL<L<LM<MH<H<EH. Taking these fuzzy sets of transactional risk into consideration, we define the boundary or range of the MARL, according to each risk propensity level as follows:

8.9.1 MARL When the Risk Propensity is Risk Averse

We consider that agents with a risk propensity level of *risk averse* accept a magnitude of transactional risk in the business activity up to the level of EL=1.

8.9.2 MARL When the Risk Propensity is Risk Neutral

We consider that agents with a risk propensity level of *risk neutral* accept a magnitude of transactional risk in the business activity up to the level of LM=1.

8.9.3 MARL When the Risk Propensity is Risk Taking

We consider that agents with a risk propensity level of *risk taking* accept a magnitude of transactional risk in the business activity up to the level of EH=1.

Using these boundaries, the risk assessing agent can evaluate the transactional risk level/s, according to his risk propensity during the risk evaluation phase. However, as mentioned earlier, it is possible that the risk propensity level of the risk assessing agent might not correspond crisply to a certain risk propensity level, but might overlap across two different adjacent levels, for example risk averse and risk neutral or risk neutral and risk taking. In such situations, the boundaries or the range of the maximum acceptable level/s of transactional risk changes from that defined earlier. This has to be measured accurately for further analysis. We discuss this further in the next section.

8.10 MARL in the Case of Overlapping Risk Propensity Levels

When the risk propensity of the risk assessing agent overlaps across two levels, the steps in determining the MARL of the risk assessing agent are:

(a) determine the Maximum Risk Propensity (MRP) of the risk assessing agent;
(b) determine the MARL of the risk assessing agent according to the MRP.

We briefly discuss these steps further in the next sub-sections.

8.10.1 *Determining the MRP in the Case of Overlapping Risk Propensity Levels*

As mentioned in the last section, the nature of the risk-taking attitude of the risk assessing agent in ascending order is: RA < RN < RT. Hence, if a risk assessing agent's risk propensity nature is a combination of levels (say RA and RN with DOM 0.4 and 0.6, respectively), then the maximum risk propensity (MRP) or maximum risk attitude of the risk assessing agent is the highest occurring risk propensity level in the order of occurrence i.e. *RN* in this particular case. The degree of occurrence of this risk propensity level is the DOM to which it quantifies to its membership function, i.e. *0.6*. Once the MRP of the risk assessing agent has been determined, then the MARL in the business activity can be ascertained, as discussed in the next sub-section

8.10.2 *Determining the MARL according to the MRP of the Risk Assessing Agent*

Depending on the MRP of the risk assessing agent, in Table 8.1, we define the fuzzy rules of the IF-THEN structure which gives the Maximum Acceptable Transactional Risk Level (MARL) in the business activity [2].

Table 8.1 Fuzzy rules to determine the MARL, according to the MRP of the risk assessing agent

	MRP		MARL
If	RA=1	then	EL=1
If	RN=0.1	then	L=0.2
If	RN=0.2	then	L=0.4
If	RN=0.3	then	L=0.6
If	RN=0.4	then	L=0.8
If	RN=0.5	then	L=1
If	RN=0.6	then	LM=0.2
If	RN=0.7	then	LM=0.4
If	RN=0.8	then	LM=0.6
If	RN=0.9	then	LM=0.8
If	RN=1	then	LM=1
If	RT=0.1	then	MH=0.34
If	RT=0.2	then	MH=0.67
If	RT=0.3	then	MH=1
If	RT=0.4	then	H=0.34
If	RT=0.5	then	H=0.67
If	RT=0.6	then	H=1
If	RT=0.7	then	EH=0.34
If	RT=0.8	then	EH=0.67
If	RT=0.9	then	EH=0.99
If	RT=1	then	EH=1

Table 8.1 lists the MARL and the DOM of the different levels of risk propensity. Once the MARL of the risk assessing agent has been determined, then the transactional level in the business activity can be evaluated. We discuss this further in the next section.

8.11 Process of Transactional Risk Evaluation in the Business Activity

Let us consider a business activity scenario between risk assessing agent 'A' and risk assessed agent 'B', where the linguistic representation of transactional risk determined is a combination of three predicates *MH=1, H=0.1818, and EH=0.0909.* While evaluating the determined level/s of transactional risk, the risk assessing agent 'A' should consider each predicate of transactional risk and then analyze whether or not it is acceptable according to his risk attitude or risk propensity. There are three possibilities, according to the risk propensity nature of the risk assessing agent 'A' as follows:

- All the levels of transactional risk in the business activity are totally acceptable to the risk assessing agent;
- All the levels of transactional risk in the business activity are totally unacceptable to the risk assessing agent;
- Some level/s of transactional risk in the business activity is/are partially acceptable to the risk assessing agent.

For example, if we consider a risk assessing agent's 'A' risk propensity nature to be RT=1, then from Table 8.1, the MARL is EH=1. Based on this, the severity levels of transactional risk (MH=1, H=0.1818, and EH=0.0909) are totally acceptable to the risk assessing agent. Similarly, if the risk propensity of the risk assessing agent 'A' is RA=1, then all the severity levels of transactional risk are beyond his MARL. When the risk assessing agent 'A's risk propensity is RN=0.6 and RT=0.4, then the MRP of the risk assessing agent is RT=0.4. In this case, he partially accepts the levels of transactional risk in the interaction with agent 'B'. In each case, the output of the risk evaluation phase and the recommended output from the fuzzy inference system to the risk assessing agent vary, according to his level of acceptance of the transactional risk severity levels in the business activity. So, while evaluating transactional risk, the risk assessing agent should consider each severity level of transactional risk with the degree of likelihood of its occurrence and then determine the impact on his risk propensity.

8.11.1 Considering the Poss Variable to Evaluate the Different Level/s of Transactional Risk in the Business Activity

In order for the fuzzy inference system to consider such cases, we introduce a variable called *Possible to Proceed in the Interaction at this stage (Poss)* when defining the fuzzy rules for transactional risk evaluation. The value for this variable is determined for each predicate of transactional risk and its purpose is twofold, namely:

(a) it represents whether or not that predicate or severity level of transactional risk, along with its magnitude of occurrence, is acceptable to the risk assessing agent, according to his risk propensity, and
(b) depending on the analysis in step (a), whether the recommended decision outcome should be *'Proceed'* or *'Don't Proceed'*.

The output which the *Poss* variable gives applies only to that predicate or severity level of transactional risk on which it is determined. If there is more than one level of severity of transactional risk determined in the business activity, then the risk assessing agent should consider each of them and determine the value for the variable *Poss* for each of them. We term the predicate or severity level of transactional risk which the risk assessing agent currently examines and determine the value of the variable *Poss* as *Current Risk Level (CRL).* By determining the value of the *Poss* variable, the effect of the risk propensity of the risk assessing agent is measuredon each predicate or severity level of transactional risk.

8.11.2 Defining the Fuzzy Inference System Rules for Transactional Risk Evaluation

The risk propensity of the risk assessing agent is represented by three fuzzy sets or predicates, as shown in Figure 8.5 and the transactional risk in the interaction is represented by six fuzzy sets or predicates, as shown in Figure 7.14. Hence, the total number of homogeneous rules to be defined for the fuzzy inference system to evaluate the transactional risk and assist in decision making is: 6 X 3 = 18. The rules that we use to define the fuzzy model are of the IF-THEN-ELSE structure. The two inputs to the fuzzy system which form the LHS of the rules are combined together with the logic 'AND' operator for the inference part to give the RHS of the rule, which is accepted if the *Poss* value is true, otherwise the 'ELSE' operator is considered. In Table 8.2, we define the rules for the system to evaluate the transactional risk and make a risk-based decision in a business activity [2].

Table 8.2 Defining the rules for the fuzzy inference system for Transactional Risk Evaluation

	CRL		CRA		RRD		Poss		RRD
If	EL	and	RA	then	P	if	1	else	DP
If	L	and	RA	then	P	if	1	else	DP
If	LM	and	RA	then	P	if	1	else	DP
If	MH	and	RA	then	P	if	1	else	DP
If	H	and	RA	then	P	if	1	else	DP
If	EH	and	RA	then	P	if	1	else	DP
If	EL	and	RN	then	P	if	1	else	DP
If	L	and	RN	then	P	if	1	else	DP

Table 8.2 (*continued*)

If	LM	and	RN	then	P	if	1	else	DP
If	MH	and	RN	then	P	if	1	else	DP
If	H	and	RN	then	P	if	1	else	DP
If	EH	and	RN	then	P	if	1	else	DP
If	EL	and	RT	then	P	if	1	else	DP
If	L	and	RT	then	P	if	1	else	DP
If	LM	and	RT	then	P	if	1	else	DP
If	MH	and	RT	then	P	if	1	else	DP
If	H	and	RT	then	P	if	1	else	DP
If	EH	and	RT	then	P	if	1	else	DP

CRA in Table 8.2 represents the fuzzy sets to which the risk propensity of the risk assessing agent quantifies to its defined membership function. The value for *Poss* in Table 8.2 for each of the CRLs being considered can be determined from Table 8.3, by taking into consideration the MRP of the risk assessing agent and then determining if the CRL is less than or equal to the MARL. If this is the case, then the value of *Poss* will be 1 for that CRL in Table 8.3, otherwise it will be 0 [2].

Table 8.3 Determining the value for the variable *Poss*

	MRP		CRL ≤		Poss		Poss
If	RA=1	and	EL=1	then	1	else	0
If	RN=0.1	and	L=0.2	then	1	else	0
If	RN=0.2	and	L=0.4	then	1	else	0
If	RN=0.3	and	L=0.6	then	1	else	0
If	RN=0.4	and	L=0.8	then	1	else	0
If	RN=0.5	and	L=1	then	1	else	0
If	RN=0.6	and	LM=0.2	then	1	else	0
If	RN=0.7	and	LM=0.4	then	1	else	0
If	RN=0.8	and	LM=0.6	then	1	else	0
If	RN=0.9	and	LM=0.8	then	1	else	0
If	RN=1	and	LM=1	then	1	else	0
If	RT=0.1	and	MH=0.34	then	1	else	0
If	RT=0.2	and	MH=0.67	then	1	else	0
If	RT=0.3	and	MH=1	then	1	else	0
If	RT=0.4	and	H=0.34	then	1	else	0
If	RT=0.5	and	H=0.67	then	1	else	0
If	RT=0.6	and	H=1	then	1	else	0
If	RT=0.7	and	EH=0.34	then	1	else	0
If	RT=0.8	and	EH=0.67	then	1	else	0
If	RT=0.9	and	EH=0.99	then	1	else	0
If	RT=1	and	EH=1	then	1	else	0

8.11.3 Defuzzifying the Output of the Fuzzy Inference System for Determination of Transactional Risk Evaluation

Once the fuzzy rules of Table 8.2 have been evaluated, they must be aggregated and defuzzified in order to obtain a crisp value on the output membership function, defined in Figure 8.6. To aggregate the output of the rules, we utilize the Root Sum Square (RSS) method. The RSS method determines the square of each rule output which corresponds to a predicate in the output membership function and then sums them to find its centroid. In our case, as there are two predicates in the output membership function, the aggregation output of all the rules for each predicate is determined by:

$$\mu \text{ `Proceed'} = \sqrt{\Sigma\,(P)2}$$

$$\mu \text{ `Don't Proceed'} = \sqrt{\Sigma\,(DP)2} \qquad (8.1)$$

The determined values for each predicate from the aggregation process are then plotted on the output membership function, to ascertain the range of the output. The scalar output of the fuzzy inference system is then obtained by defuzzifying the range in which the output exists to obtain a crisp value by utilizing the center of gravity or centroid method. The obtained crisp value, when plotted on the output fuzzy set, represents the evaluation of the transactional risk and the recommended risk-based decision from the fuzzy inference model.

In the next section, we will summarize and also demonstrate with an example the working of the fuzzy inference system in assisting the risk assessing agent to evaluate transactional risk and determine it for a business activity.

8.12 Scheme of Operation of the Fuzzy Inference System for Transactional Risk Evaluation

The operation of the fuzzy inference system in evaluating transactional risk can be classified into two phases, namely the learning and the computation phase. In the learning phase, the fuzzy inference system 'learns' about the tendency of the risk assessing agent in taking risk in the business activity, and in the computation phase, it utilizes the determined tendency of the risk assessing agent in computing the output of the risk evaluation and the recommended risk-based decision in forming a business activity.

8.12.1 Learning Phase

During the learning phase, the risk assessing agent enters his risk propensity level on a scale of 0-5, which the fuzzy inference system converts and then determines the DOM of each defined risk attitude, according to the membership function defined in Figure 8.5. As mentioned earlier, with varying severity level/s of risk attitude of a risk assessing agent, his tendency to take the level of risk in the interaction also changes. So, the fuzzy inference system has to consider this characteristic while evaluating the severity level of transactional risk. This is achieved by the fuzzy inference system, which utilizes the determined DOM of the risk propensity level/s of the risk assessing agent to determine his MRP and, based on this, ascertains the MARL. This is done by the fuzzy rules defined to ascertain them, as explained in Table 8.1.

The schema of the operation of the fuzzy inference system in this phase is as follows:

- It takes as input the risk propensity nature value of the risk assessing agent from its UoD.
- Based on this value, it then quantifies and determines the DOM of each fuzzy set from the spectrum, which represents the risk-taking nature or the risk attitude of the risk assessing agent. In doing so, the fuzzy inference system accurately ascertains the risk taking nature or the risk propensity of the risk assessing agent in the business activity, from its available risk attitudes.
- Based on the DOM determined for each category of risk attitude, the fuzzy system then ascertains the Maximum Risk Propensity (MRP) of the risk assessing agent. In other words, the MRP of the risk assessing agent represents the highest level of risk attitude of the risk assessing agent along with its DOM in the interaction.
- The fuzzy system then determines the Maximum Acceptable Transactional Risk Level (MARL) of the risk assessing agent, based on his MRP in the interaction. Once the fuzzy system has completed learning about the tendency of the risk assessing agent, then it starts the computation phase.

8.12.2 Computation Phase

In this phase, each predicate or severity level of transactional risk determined in forming a business activity with a risk assessed agent along with its DOM is evaluated to ascertain whether it is acceptable to the risk assessing agent or not, according to its MARL. Based on this evaluation, a value of *Poss* for that specific predicate is determined, according to the rules explained in Table 8.3. When all the level/s of transactional risk are evaluated, the fuzzy rules defined in Table 8.2 are used to ascertain the output of the risk evaluation phase and the recommended risk-based decision to the risk assessing agent in forming an interaction with a risk assessed agent.

The schema of the operation of the fuzzy inference system in this phase is as follows:

- The determined linguistic level/s of transactional risk in forming the business activity are considered.
- The fuzzy system then considers each severity level of transactional risk, which is termed the CRL, along with its magnitude and then determines the value of the variable *Poss* for that specific level of transactional risk. The variable *Poss* stands for *Possible to Proceed in the Interaction at this stage*. This variable represents whether the considered level of transactional risk

(CRL), along with its magnitude of occurrence, is acceptable or not to the risk assessing agent, according to its MARL in the interaction, and based on this, whether the risk assessing agent at this stage can proceed with the risk assessed agent in an interaction. The fuzzy system determines the value for the variable *Poss* for each level or predicate of transactional risk while interacting with the risk assessed agent.

- Once the MRP, MARL and Poss values for each severity level or predicate of transactional risk have been determined, then the fuzzy inference system evaluates the fuzzy rules in order to determine the strength to which the output predicate of each rule fires.
- The inference system then aggregates the output from the rules corresponding to each fuzzy predicate on the output membership function by using the Root Sum Square method.
- The determined strength of each output fuzzy predicate is then defuzzified to obtain a crisp scalar value on the output membership function.
- The scalar value, when plotted on the output membership function, represents the recommended output of the fuzzy inference model to the risk assessing agent on forming an interaction with the risk assessed agent.

If the risk assessing agent has to decide on an interaction when there is more than one risk assessed agent, the fuzzy inference system repeats the above steps and determines the scalar output on the output membership function in forming an association with each risk assessed agent. This enables the risk assessing agent to evaluate the transactional risk in forming a business activity with each agent and make an informed decision on an interaction.

8.13 Example of Transactional Risk Evaluation in a Business Activity

To explain the operation and effectiveness of the proposed fuzzy inference system while undertaking transactional risk evaluation, let us consider the interaction scenario discussed in Chapter 5, where a risk assessing agent 'A' wants to interact with a logistics company to transport its goods. Let us consider that the risk assessing agent 'A' has to choose from three logistics companies (which we term agent 'B, agent 'D' and agent 'E', respectively) to form a business association. Let us consider that the linguistic representations of transactional risk in forming a business activity with each agent are as follows:

- ➢ Linguistic transactional risk in interacting with agent 'B':
  MH=1, H=0.1818, EH=0.0909.
- ➢ Linguistic transactional risk in interacting with agent 'D':
  MH=1, H=0.9, EH=0.5.
- ➢ Linguistic transactional risk in interacting with agent 'E':
  EL=1, L=1, LM=0.1.

To evaluate transactional risk, risk assessing agent 'A' has to first determine his risk propensity nature on the membership function, shown in Figure 8.5. Let us consider that the risk assessing agent chooses a value of 3.6 as his Risk Propensity in the interaction. The DOM of the predicates or fuzzy sets to which the Risk Propensity of the risk assessing agent quantifies is a combination of two different risk propensity fuzzy sets, *Risk Neutral and Risk Taking* with a DOM of:

$\mu_{RP}(RN) = 0.70$ and $\mu_{RP}(RT) = 0.30$

8.13.1 Recommended Risk-Based Decision on Interaction with Agent 'B'

According to the risk propensity of the risk assessing agent, his MRP is RT=0.3. From Table 8.1, the *Maximum Acceptable Transactional Risk Level(MARL),* based on this degree of risk propensity, is MH=1.0. The fuzzy inference system considers each predicate of transactional risk (CRL) and determines if it is within the MARL of the risk assessing agent, according to his risk propensity. Based on its computation, the value for the variable *Poss* is determined for each predicate, and by using the rules defined in Table 8.2, it determines the strength to which each rule fires and the DOM of each output rule. The aggregated DOM of the predicates or the fuzzy sets of the output variable determined by using equation 8.1 are:

$\mu_{RRD}(P) = 0.761$ and $\mu_{RRD}(DP) = 0.287$

Once the DOM of each output predicate has been determined, then it must be plotted on the output membership function to determine the range in which the scalar output of the system exists. Figure 8.7 shows the range of the output membership function based on the DOM of each output predicate. The scalar output of the fuzzy inference model, obtained by defuzzifying the shaded region, is 7.369. By plotting the scalar output on the output membership function, the DOM of the output fuzzy set *Proceed* is 0.7369, whereas that of the fuzzy set *Don't Proceed* is 0.2631. In other words, this implies that, based on the risk propensity of the risk assessing agent and the severity level/s of transactional risk in the business activity with 'B', the strength to which the rules fire and quantify to the output fuzzy set *Proceed* and *Don't Proceed* are 73.69% vs 26.31%, respectively.

8.13.2 Recommended Risk-Based Decision on Interaction with Agent 'D'

While making a decision, in order for the risk assessing agent to avoid being biased towards a particular agent, and to assess all the risk assessed agents of the

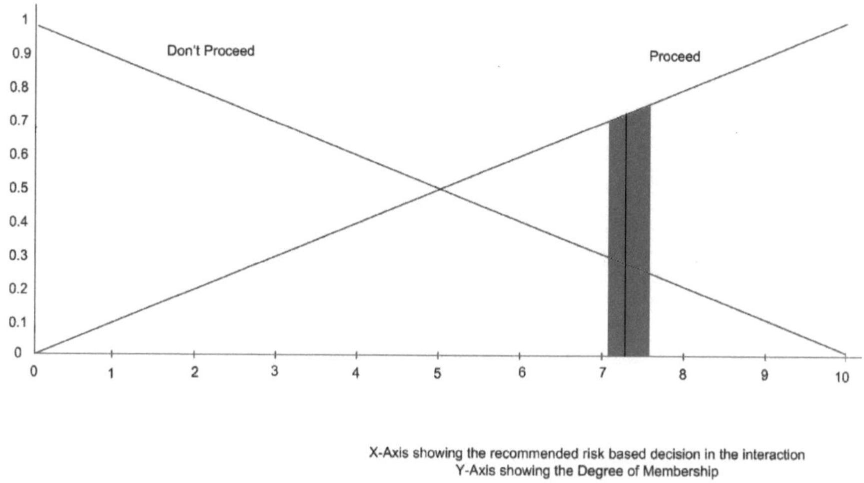

Fig. 8.7 The range of the output fuzzy set in interacting with agent 'B'

interaction in a uniform way, we propose that his risk propensity remains constant throughout the process of choosing an agent with which to interact from a set of risk assessed agents. Subsequently, the risk propensity nature of the risk assessing agent 'A', while evaluating transactional risk in forming a business activity with agent 'D', is the same as it was while assessing agent 'B'. The MRP and MARL of the risk assessing agent 'A' remains same as the previous scenario, as his risk propensity is unchanged. The aggregated DOM of the predicates or the fuzzy sets of the output variable determined by using equation 8.1 are:

$\mu_{\text{RRD}}(\text{P}) = 0.761$ and $\mu_{\text{RRD}}(\text{DP}) = 0.959$

Figure 8.8 shows the range to which the output function extends, based on the DOM of its predicates. The scalar output of the fuzzy inference model obtained by defuzzifying the shaded region is 4.004. By plotting this value on the output membership function, the DOM of the output fuzzy set *Proceed* is 0.4, whereas that of the fuzzy set *Don't Proceed* is 0.6. In other words, this implies that, based on the risk propensity of the risk assessing agent and the severity levels of transactional risk in interacting with the risk assessed agent 'D', the strength with which the rules fire and quantify to the output fuzzy set *Proceed* and *Don't Proceed* are 40% vs 60%, respectively.

It can be seen that the severity level or predicates of transactional risk in interacting with agents 'B' and 'D' are the same, but the likelihood of occurrence of each level differs. As the likelihood of occurrence of an unacceptable severity level of transactional risk, according to the risk attitude of the risk assessing agent,

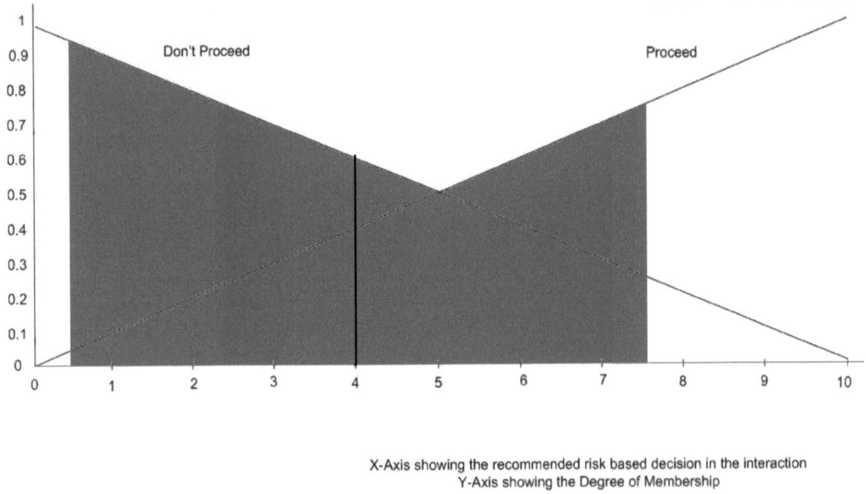

Fig. 8.8 The range of the output fuzzy set in interacting with agent 'D'

is greater in the case of agent 'D' compared to agent 'B', the strength by which the predicates of the output membership function quantifies also vary accordingly, in interacting with both agents. In the next section, we will determine the fuzzy recommended risk-based decision output on an interaction with the risk assessed agent 'E'.

8.13.3 Recommended Risk-Based Decision on an Interaction with Agent 'E'

To determine the impact of the risk propensity of the risk assessing agent on each predicate of the transactional risk in interacting with agent 'E', the aggregated DOM of the output fuzzy sets are:

$\mu_{RRD}(P) = 1.0$ and $\mu_{RRD}(DP) = 0$

In this case, there is no range as both the output variables are on the same point on the output membership function. By plotting the scalar output on the output membership function, the DOM of the output fuzzy set *Proceed* is 1.0, whereas that of the fuzzy set *Don't Proceed* is 0.0. In other words, this implies that, based on the risk propensity of the risk assessing agent and the level and magnitude of transactional risk in interacting with agent 'D', the strength to which the rules fire and quantify to the fuzzy output fuzzy sets *Proceed* and *Don't Proceed* are 100% vs 0% respectively. Figure 8.9 shows the scalar output on the output function.

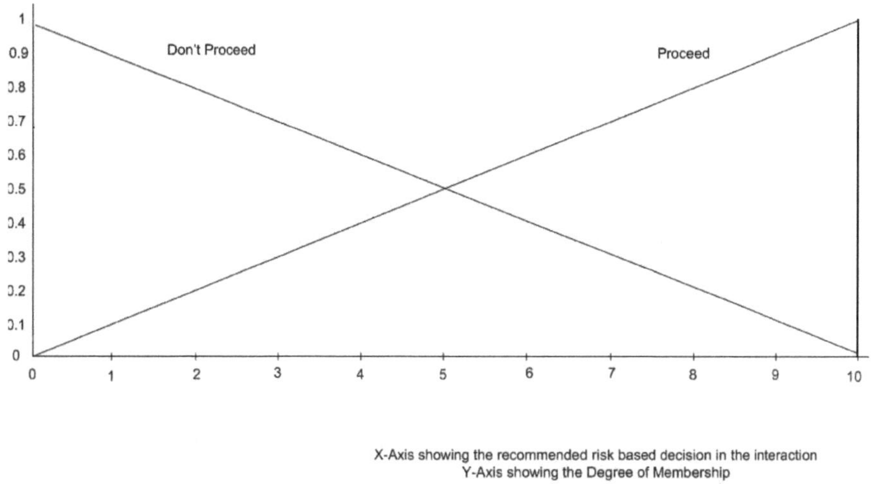

Fig. 8.9 The range of the output fuzzy set in interacting with agent 'E'

8.13.4 Observations of the Transactional Risk Evaluation Phase

From the example discussed above the following conclusions can be drawn:

- The strength to which the rules correspond to the fuzzy set *Proceed* are 73.69%, 40% and 100% while interacting with agent 'B', 'D' and 'E', respectively.
- Of the set of risk assessed agents, it can be concluded that the fuzzy inference system recommends agent 'E' to the risk assessing agent 'A', as the strength to which the rules fire and produce an output that leans towards the output fuzzy set 'Proceed' more while interacting with the other agents.
- An alternative argument for the recommended decision by the fuzzy inference system to choose agent 'E' is that the strength to which the rules fire and produce an output leans towards the output fuzzy set *Don't Proceed* less in the case of agent 'E' than with agents 'B and 'D': 0% as compared to 26.31% and 60%, respectively.
- The degree to which the output fuzzy sets *Proceed* and *Don't Proceed* quantifies in interacting with an agent are, according to the predicate/s or severity level/s of the transactional risk and its/their possibility of occurrence, present in interacting with that agent, which is/are then further classified depending on whether it is/they are acceptable or unacceptable to the risk assessing agent, according to his risk propensity or risk attitude.
- In the example considered, it can be seen that even though the same rules fire in the case of agents 'B' and 'D', the strength to which each rule quantifies to its output set varies. This is due to differences in the level of occurrence of the predicates of transactional risk in interacting with both the agents, which varies the output of the fuzzy inference system, accordingly.

- Furthermore, it can be seen that the output from the fuzzy inference model is neither 100% Proceed nor 100% Don't Proceed in the case of agents 'B' and 'D', as the severity levels of transactional risk along with their magnitude of occurrence in interacting with the risk assessing agents are not either completely unacceptable or completely acceptable to the risk assessing agent, as in the case in interacting with agent 'E'.
- While aggregating the fuzzy rules in interacting with agent 'D', it can be seen that if the fuzzy additive method had been used, then the output fuzzy sets would have a DOM of 1.0 each. This would give the resultant answer from the fuzzy inference system as 0.5 *Proceed* and 0.5 *Don't Proceed*. The disadvantage of the max-min method, discussed in the previous chapter, applies in this case, too. Hence, these aggregation techniques were not used in this inference model and the RSS aggregation approach was used, which determines the strength of the output fuzzy set by considering all the firing rules.
- By using the fuzzy inference system, the risk attitude of the risk assessing agent is accurately determined, even if it spreads across different possible risk propensity levels. Based on the accurate risk propensity level/s of the risk assessing agent, its effect on transactional risk in interacting with a risk assessed agent, and hence the decision output, is determined accurately.

8.14 Recapitulation

In this chapter, we proposed a fuzzy inference system which assists the risk assessing agent to evaluate the determined severity level/s of transactional risk and make an informed decision about an interaction with a risk assessed agent. The main purposes of using a fuzzy inference system to assist the risk assessing agent in determining on an interaction are:

- In contrast to the other approaches in the literature, the fuzzy inference system allows the risk assessing agent to consider the intermediate states from the spectrum of the risk propensity levels as its risk taking attitude in the interaction.
- It considers each severity level or predicate of the determined transactional risk in forming an interaction with a risk assessed agent, and ascertains the effect of its risk propensity on it, to determine the recommended risk-based decision in undertaking an interaction with that agent.

Furthermore, the output which the fuzzy inference model gives is not just limited to an output predicate, such as either *Proceed* or *Don't Proceed,* rather it also gives the strength to which each of the output fuzzy set quantifies. This is particularly useful for the risk assessing agent when he has to decide on a business activity. Based on the evaluated level/s of transactional risk, the next phase in the process of risk analysis is Transactional Risk Management. We discuss this further in the next chapter.

References

1. Sitkin, S.B., Weingart, L.R.: Determinants of Risky Decision-Making Behavior: A Test of the Mediating Role of Risk Perceptions and Propensity. Academy of Management Journal 38, 1573–1592 (1995)
2. Hussain, O.K., Dillon, T.S., Hussain, F.K., Chang, E.: Transactional Risk-based Decision Making System in E-Business Computing. Computer Systems Science and Engineering 25, 15–25 (2010)

Chapter 9
Risk Management Phase: Transactional Risk Management in Business Activities

9.1 Introduction

In the previous chapters, we proposed an approach by which transactional risk in a business activity can be ascertained and evaluated against the risk propensity of the risk assessing agent. The next phase in the process of risk analysis is risk management, where appropriate steps are taken to address those risks in the business activity which have been evaluated as being beyond the acceptable level. However, as mentioned in Chapter 3, it is important to note that the process of risk management is business activity-specific and it is not possible to define a single process that can be generalized to every business activity. In other words, the goals, objectives and desired outcomes of each business activity vary. Accordingly, the risk treatment steps taken to address the unacceptable level/s of transactional risk also vary as they depend on:

(a) the severity of the determined level/s of transactional risk in the business activity; and
(b) the goals to be achieved in the business activity.

As a result, it is not possible to have a generalized framework that lists all the risk treatment steps for transactional risk management in a business activity. Rather, the steps have to be interaction and goal specific. But, it is possible to have a generalized framework which measures the adjusted transactional risk in the business activity as a result of the outcomes of the interaction-specific risk treatment steps taken. This generalized framework measures transactional risk by using the approaches proposed in the earlier chapters. In this chapter, we explain how such a generalized framework can be used for transactional risk assessment and management by applying the approaches presented in the previous chapters in a case study of just-in-time production [1, 2].

O. Hussain et al.: Risk Assessment & Management in the Networked Economy, SCI 412, pp. 243–261.
springerlink.com

9.2 Case Study for Transactional Risk Management: Just-in-Time Production

9.2.1 Expectations in Just-in-Time Production

In Demand-Driven Production, the basic mandatory requirement to which manufacturers commit is the Just-in-Time (JIT) concept or time-based manufacturing [3], to ensure that goods are manufactured, produced and assembled by the manufacturing company, according to the given specifications of the order received from the customer within the given time. The manufacturing company can either produce the goods in its own manufacturing plant(s), which we term in-house production; contract the production to a third party manufacturing company, which we term out-sourced production; or utilize both these types of production to obtain the final end product(s), according to the customers' order and specifications, within the specified time frame. But, whatever the production mechanism utilized, the manufacturer must ensure that the output generated by its manufacturing plant(s) must be proportional to the real-time demand of the customers. So, the *desired outcomes* in this particular business scenario are:

(a) The customers expect the manufacturer to commit to their order and deliver the goods in time.
(b) The manufacturer expects they will be able to meet the customers' order on time, thereby ensuring they contain the financial cost of production and receive financial revenue from the customer, which is important for the company's sustainability and economic growth.

In this case study, we look at the problem of transactional risk assessment and management from the viewpoint of the manufacturer. So, the *expectation* of the particular business activity is defined as follows [1]:

In order for a demand-driven manufacturing company to produce and deliver customized goods to the consumers within the specified period of time and cover its operational costs plus generate profits, it has a target to achieve in terms of revenue return relating to the specific levels of product/s demand.

The question that we want to analyze and manage in this chapter is:

What are the risks which may lead to the possibility of the manufacturer not achieving their target and, based on the analysis, what steps can it take to manage this risk?

9.2.2 Identification of Uncertainties That May Lead to the Non-achievement of the Expectations

In the previous sub-section, we defined the expectations of the business activity. However, due to various factors/uncertainties, the manufacturer might not be able to produce the required end product/s as desired by the customers within the specified period of time. This will have serious implications for the manufacturer, both financially and in terms of its reputation. To avoid the occurrence of such a scenario, various uncertainties that may result in this should be identified and appropriate steps taken to mitigate the risk arising from them. During transactional risk analysis, we are interested in identifying the uncertainties that may lead to the failure to meetthe expectations in the business activity. In this case study, we take three sources of uncertainty into account [1], namely:

(a) uncertainty in demand over different time periods in a given time;
(b) uncertainty in the failure rate/s of the production units involved in production; and
(c) uncertainty in the availability of inputs if a Just-in-Time (JIT) approach to input raw materials or components is adopted.

Using these uncertainties, in this chapter, we aim to determine the transactional risk of the manufacturer in not meeting the expected demand which is measured by the following outcomes:

- the probability of the manufacturer not meeting the required demand (measured in financial terms) within a given time period;
- the unserved demand (measured in financial terms) by the manufacturers' production units; and
- the degree of loss to the manufacturer (measured in financial terms) as a result of unserved demand.

The solution to the problem will be of primary interest to manufacturers as it will help them make accurate decisions about their machines' capacity and the customer demand that they can expect to meet in relation to their production capabilities. This production capacity should be planned carefully and production equipment should be purchased which matches the anticipated future demand of customers. In the next section, we define several key terms which we use in this case study for transactional risk assessment and management.

9.3 Definition of Key Terms in Transactional Risk Management: Just-in-Time Production

9.3.1 Demand

In the previous sub-section above, we noted that the uncertainty of *demand* over different periods of time has to be identified. This demand relates to the different goods to be produced by the manufacturer which generate a certain amount of revenue. We refer to this demand as the revenue generated by customer orders. In other words, we characterize demand of a time period as the total revenue that would be generated by fulfilling <u>all</u> customer orders in the required time frame. This ensures that all the required product(s) will be delivered to the consumer(s) in the specified period of time.

9.3.2 Demand Expected Curve (DEC)

Demand Expected Curve (DEC) measures the variability in demand over a period of time. Similar to the AIC discussed in Chapter 6, the DEC represents the probability of an amount of revenue to be generated by the manufacturer in a given time period.

9.3.3 Forced Outage Curve (FOR)

Similar to the FLC and VC discussed in Chapters 5 and 6 respectively, *Forced Outage Rate (FOR)* is the probabilistic function that shows the different severity level/s of outage and the magnitude of occurrence of outage level/s of a production unit from the desired operating capacity.

9.3.4 Uncertainty in the Availability of Raw Materials Curve (URMC)

Uncertainty in the availability of raw materials Curve (URMC) is a probabilistic curve that represents the different severity level/s of non-availability and the magnitude of occurrence of those level/s in the required raw material in a given time period.

9.3.5 Equivalent Demand Expected Curve (EDEC)

The forced outage rates of the different production units and the uncertainty in the availability of the raw materials will invariably have an effect on the expected demand that has to be met within a time period. In order to determine the effect, we calculate the *Equivalent Demand Expected Curve (EDEC)* of the time period. This is similar to the TFAIC described in Chapter 6.

9.3.6 Maximum Achieved Demand (MAD)

Maximum Achieved Demand (MAD) represents the demand (in financial amounts) achieved by the manufacturer's production units in a given period of time. This is similar to the MIC discussed in Chapter 6.

9.3.7 Loss of Demand Probability (LODP)

Loss of Demand Probability (LODP) represents the probability of the manufacturer not meeting the required demand from its production units in the specified time frame. This is similar to the LOIP discussed in Chapter 6.

9.3.8 Unserved Demand Curve (USDC)

Unserved Demand Curve (USDC) represents the cumulative probability of the unserved demand in the time period. This demand is the part of the EDEC curve after the MAD. USDC is similar to the LC described in Chapter 6.

9.3.9 Expected Loss Curve (ELC)

Expected Loss Curve (ELC) represents the probabilistic curve that represents the unserved demand by the manufacturer in a given time period, due to the occurrence of various uncertainties. This is similar to the LC discussed in Chapter 6.

9.3.10 Cost Hedging Curve (CHC)

Cost Hedging Curve (CHC) represents the cost of meeting the unserved demand by the manufacturer's production units over the time period.

Using these terms, in the next section, we briefly discuss the process of how transactional risk can be assessed and then managed in the particular business activity.

9.4 Transactional Risk Assessment: Probabilistic Assessment of Loss in Revenue Generation in Just-in-Time Production

9.4.1 Identifying Demand over a Given Time Period

One of the pre-requisites for manufacturers in Just-in-Time production is to forecast the expected demand of its customers for different products, over a given period of time. It is likely that demand will fluctuate over a given period of time and may not follow the patterns of previously experienced scenarios. For example, there may be elements of stochastic variation, trends in variation or seasonality in variation. So, in order to measure such variability in demand, we aim to carry out a probabilistic determination over a given period of time. We term the total period of time in which the demand has to be determined as the 'time space'. The time space is broken down into different equal sub-intervals of time (say a week or month) that are termed 'time slots'. The expected demand for each time slot might vary, according to the real time customers' orders and their corresponding operational costs. The manufacturer should measure this variation while determining the demand in that time slot.

For example, let us consider that a manufacturer wants to determine the risk of it not meeting the expected demand within a time period. The manufacturer decides that there are five time slops in the particular time space. Let us assume that the expected demand is in the order of \$500, \$2,000, \$2,500, \$2,000 and \$3,000 in time slots 't1', 't2', 't3', 't4' and 't5' of the time space, as shown in Figure 9.1 [1].

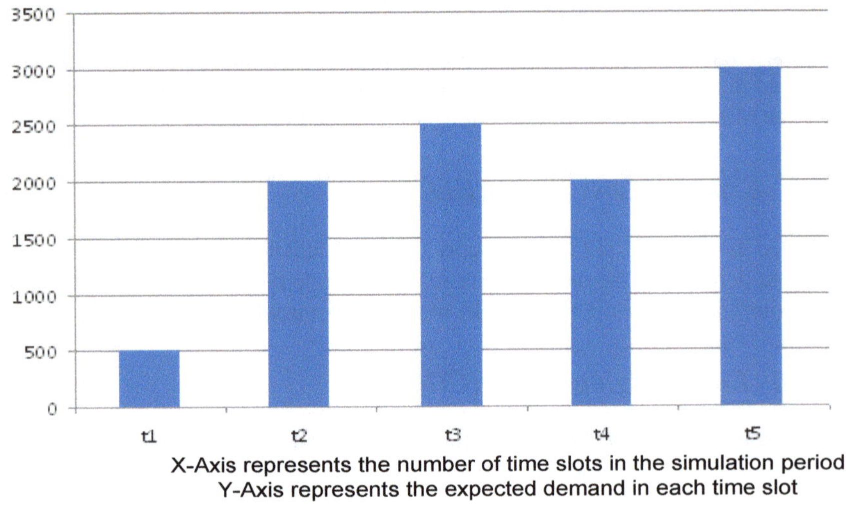

Fig. 9.1 The expected demand in each time slot of the time space

From the expected demand over the time period, the *Demand Expected Curve* (DEC), as shown in Figure 9.2, is plotted. As mentioned earlier, the DEC represents the probability of an amount of revenue to be generated by the manufacturer in each time slot of the time space.

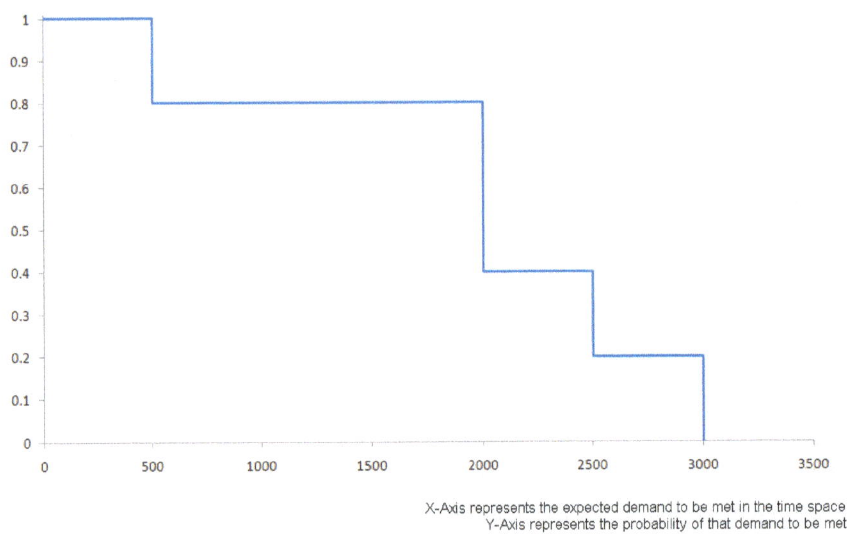

Fig. 9.2 The DEC according to the expected demand in the time period

The manufacturer meeting the demand represented by the DEC would signify that:

a) it delivers the customers' orders on time and meets its operational costs; and
b) its manufacturing units produce the equivalent worth of goods in accordance with the expected demand within the time period.

9.4.2 Measuring the Different Severity Level/s of Occurrence in Uncertainties in the Business Activity (FOR and URMC)

As discussed earlier, there may be several uncertainties that could result in the manufacturer not meeting the expectations as desired. In this case study, we consider two such uncertainties, namely Forced Outages of Production Units (FOR) and Uncertainty in the availability of Raw Materials (URMC). FOR of a production unit represents that production unit not operating to its full expected capacity, thus affecting the planned output, which in turn affects the production of goods and the expected output to be generated from it. URMC represents the level of raw materials not available in the required quantity at the required period of time. If this situation occurs, then even if the production units are working to full capability, the manufacturer will not be able to produce the required demand, which will result in it not meeting the required demand in a given time period. This will increase the demand to be met in the next time slot, in addition to the normal expected demand.

To understand the effect of these uncertainties, their severity in the level of occurrence and magnitude of the level of occurrences from the desired point throughout the time period of the business activity should be ascertained. This is achieved by using a probabilistic curve that represents the different level/s of variation in the occurrence of these variables and the magnitude of occurrence.

Approaches in the literature utilize a two-state probability function to represent the variation of occurrence of an uncertain variable. This is shown in Figure 9.3 which represents either the occurrence of a variable at an optimal point with a probability 'p', or not occurring at the optimal point with a probability 'q = 1-p' [4] (for example a production unit either working at desired capacity with probability 'p' or not working with probability '1-p').

However, in the real world, it is highly possible that there might be different level/s or degree/s of variation from the desired point of each uncertain variable 'C'. For example, even if the production unit is not working at the desired optimal level, it may still be working at a de-rated or lower capacity, as shown in Figure 9.4 [1].The different level/s or de-rated degree of operation of a manufacturing unit might not necessarily cause its production process to shut down completely, but it represents de-rated operating states that may prevent it from reaching its desired maximum production capacity for the time period of interest. This will result in that production unit's output being below the required desired capacity, as shown in Figure 9.5, therefore it will not meet the demand that is expected from it in that time period [1].

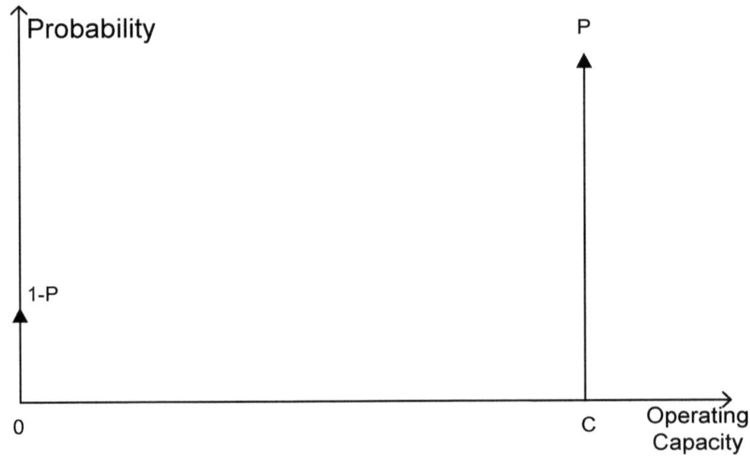

Fig. 9.3 A two-state discrete random representation in the occurrence of a variable in a time slot

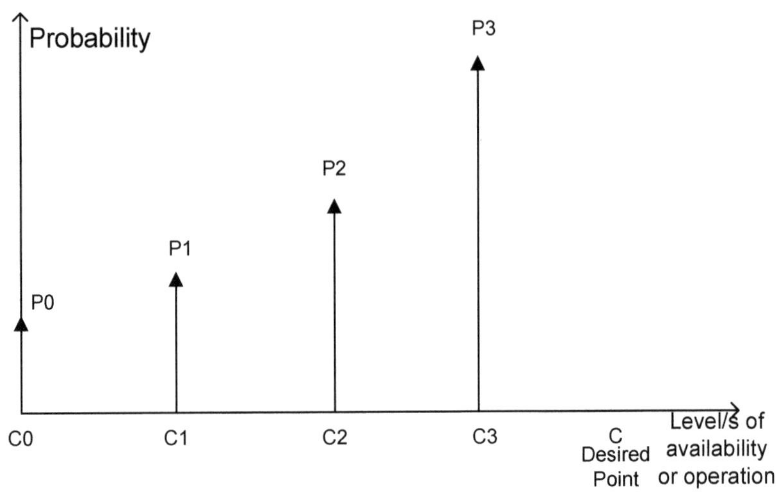

Fig. 9.4 Multi-state discrete density function showing the different level/s of operation or availability of a variable from desired point 'C'

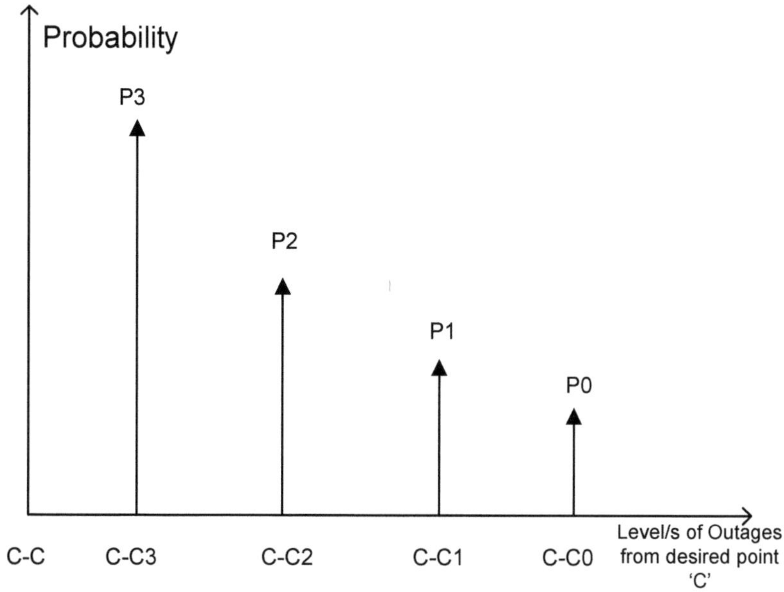

Fig. 9.5 Forced Outage Rates in the operation or availability of a variable from desired point 'C'

The situation of uncertainty in the availability of raw materials is similar. These level/s of de-rated operations of production units or availability in the availability of raw materials are different from the two-state representation shown in Figure 9.3, and thus have to be identified appropriately while ascertaining the impact on meeting the required demand (DEC) of the given time period. We propose that the manufacturer ascertains the different level/s of outages of the operation or the availability of each variable from desired point 'C' and models it using the representation shown in Figure 9.5.

9.4.3 Ascertaining the EDEC of the Business Activity

After determining the multi-level/s of deviation in the occurrence of uncertain variables from the desired points (FOR and URMC), the EDEC of the business activity should be ascertained. The EDEC, as shown in Figure 9.6, represents the extra financial amount that has to be met by the manufacturer, taking into account variations in the occurrence of the uncertain variables from the desired points [2].

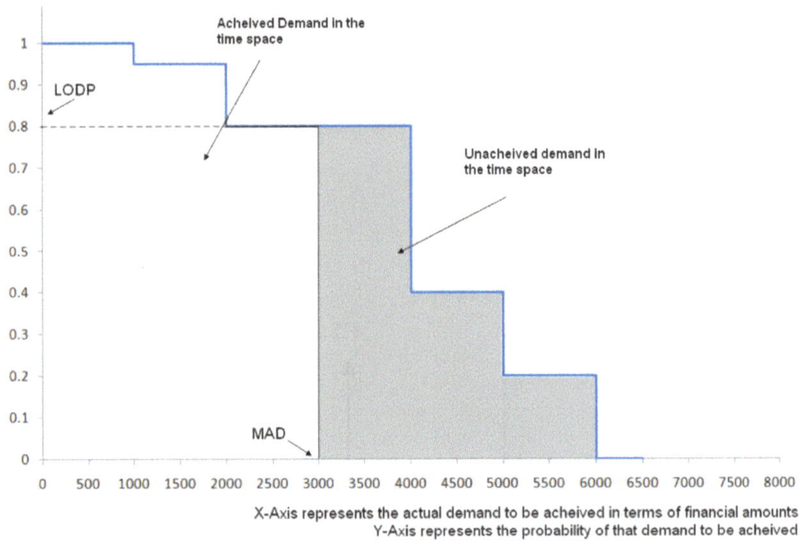

Fig. 9.6 The EDEC for the duration of the business activity

Compared to the DEC, the EDEC shows an increased probability in the amount of expected demand to be met in the time space by considering the different types of uncertainties. This modification reflects the fact that when some production unit(s) are randomly forced offline, the remaining production unit(s) should see a backlog of goods to be produced and should compensate for the demand that is being unserved. Similarly, when there is a shortfall in some raw materials, then the backlog and effect of that shortfall in the next time slot is determined and taken into consideration. To determine the financial loss due to such uncertainties, the manufacturer should ascertain the:

- Loss of Demand Probability (LODP);
- Unserved demand Curve (USDC) of the given time period.

9.4.4 Ascertaining the LODP and USDC in the Business Activity

The LODP index of a time period, as shown in Figure 9.6, is the ordinate on the EDEC corresponding to the maximum amount of generated revenue (MAD) by the manufacturer on the abscissa. The USDC of the time space represents the additional demand which has to be generated and met by the manufacturer from the MAD of its production units. But this level(s) of demand will not be met by the manufacturer in the time space in the existing configuration, as it is beyond the capacity of its production unit(s) and can be termed as the financial loss that it will experience, if appropriate strategies are not taken to address this. This financial loss can be termed as the transactional risk to the manufacturer in the business activity and he should develop management plans to mitigate such risks.

9.4.5 Strategies to Manage Transactional Risk in the Business Activity

To minimize such losses, the manufacturer should develop strategies that allow it to meet the unserved demand in the time space. One such strategy is to anticipate production unit outages or a shortage in raw materials and ensure there is a stockpile of vital parts or materials on hand, according to a set of specifications in its inventory. However, this may lead to inventory paradox where the manufacturer does not clear the stored stock in a timely manner. An example of this is the Amdahl Computer Corporation which held an inventory of all possible configurations, leading to hundreds, if not millions of dollars of excess inventory [5].

Another strategy which the manufacturer can adopt to mitigate this type of risk is to hedge the production of the required unserved demand with third party producers, hence meeting the total expected demand in the time space. In this case study, by hedging we mean the process of utilizing third party producers for manufacturing the goods and paying them for their services. But, before utilizing this strategy, a comparative analysis has to be carried out to determine whether it is feasible or profitable for the manufacturer to hedge with third party producers. Such an analysis is important and will have a significant impact on its decision-making. In this chapter, we utilize such an approach for risk management and propose a methodology for undertaking a comparative analysis to assist decision-making, which, if beneficial, can be used as a risk treatment and management strategy in the specific business activity.

9.5 Comparative Analysis in Hedging with Third-Party Producers When Unserved Demand Is Represented as a Crisp Financial Amount

By undertaking a comparative analysis, we aim to analyze and compare the cost that will be incurred by the manufacturer in hedging goods from third-party producers with the loss that would be incurred by not meeting the required demand of the given time period. An important point to consider here is that such an analysis has to be done only on the unserved demand of the time period, which is the part of the EDEC beyond the MAD, as shown in Figure 9.6. The series of steps to be followed are [2]:

(a) Taking the part of the EDEC after the MAD, the manufacturer should first determine the crisp financial amount unserved demand of the time space.
(b) Determine the cost of hedging the unserved demand from third party producers and utilize the analysis to make a decision.

We discuss these steps further in the next sub-sections.

9.5.1 Determine the Crisp Financial Amounts of Unserved Demand

The probability mass function of the EDEC beyond the MAD should be determined, which will give the actual probability of each amount which has to be generated by the production unit(s) in the time space of the business activity. To calculate the unserved demand of the time space, the manufacturer should consider the part of the EDEC after the MAD and utilize the probability of each unserved amount to determine the number of time slots in which that amount had to be achieved in the time space. Based on the results of the analysis, the manufacturer can ascertain the unserved demand as a crisp financial amount in the time space.

For example, let us consider the EDEC as shown in Figure 9.6 where the MAD in the time space is $3,000. In order to decide whether to hedge with third-party producer(s) to meet the total demand, the manufacturer needs to first determine as a crisp amount the level of unserved demand of the time space. The probability mass function of the EDEC is shown in Figure 9.7 as follows [2]:

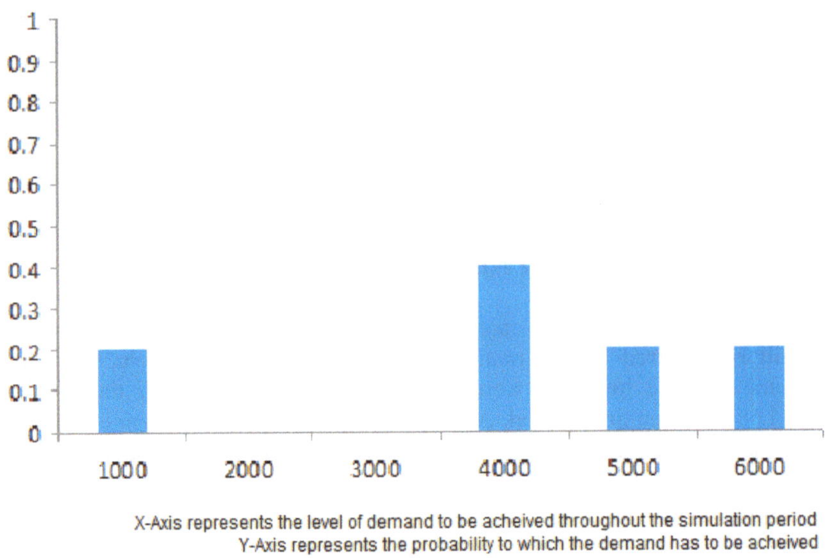

Fig. 9.7 The Probability Mass Function of the EDEC

To determine the crisp unserved financial amount, the manufacturer should determine the number of time slots where such level of loss is experienced by utilizing the probability of each amount on the EDEC after the MAD. As mentioned in Section 9.4.1, in the current example, we consider that there are 5 time slots in the time space. Figure 9.8 represents the occurrence of each level of unserved demand from the MAD (or the loss that can be experienced) in the number of time slots of the time space. From Figure 9.8, it can be seen that the crisp amount of loss that could be incurred by the manufacturer during the time space is $7,000.

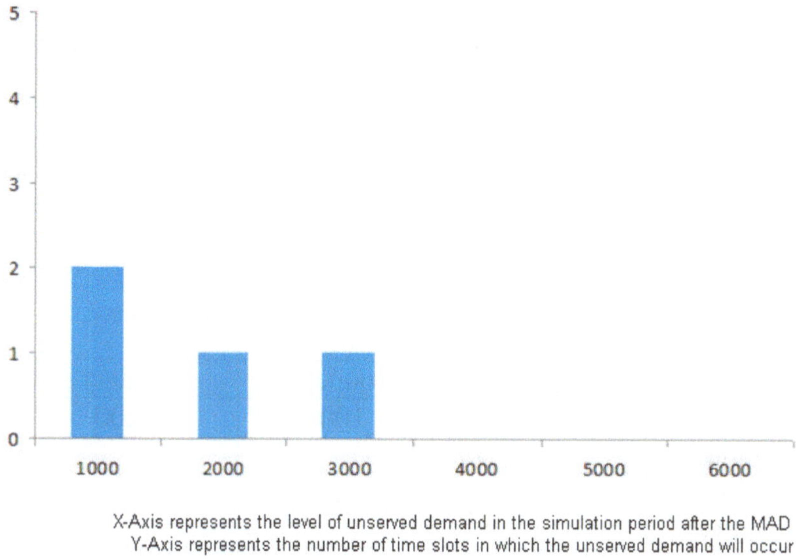

Fig 9.8 Occurrence of the different level/s of unserved demand in the time slots of the time space

9.5.2 Determine the Cost of Hedging the Unserved Demand from Third-Party Producers

By considering the downtime of the manufacturer's production units and uncertainty in the availability of the raw materials, the manufacturer can determine the type of products and their quantity that will form the unserved demand of the time period. Utilizing such information, the manufacturer will be able to determine the costs that it would incur in hedging those products with third-party producers. It can then undertake a comparative analysis of the costs of hedging the production of the goods with the loss that it would otherwise experience by not meeting the demand, and then decide whether to consider this option to manage the identified risks. If the cost of hedging the goods from third-party producers is less than the loss that it would incur by not meeting the customer's demand, then the manufacturer can consider this strategy in order to meet its total demand in the time space. On the other hand, if the analysis results show that the loss the manufacturer would incur is less than what it would spend to hedge the goods from third-party producers, then it need not consider this option in order to meet the expected demand unless supplying the requested goods to a customer has high strategic business significance e.g. a very large and important customer with which the manufacturer would not like to damage its reputation.

Continuing the example scenario discussed earlier, let us consider that the manufacturer, by analyzing the operational history of its production units and uncertainty in the availability of raw materials, determines that it has to hedge the production of *Product 1*, *Product 2* and *Product 3* from third-party producers in order to meet the total demand. Let us consider that the manufacturing and operational costs of the third-party producer depend on the quantity of each product to be manufactured, as shown in Table 9.1 [2].

Table 9.1 Operational Costs of manufacturing goods from a third-party producer

	Product 1	Product 2	Product 3
Quantity required	1000	100	200
Cost per quantity	$ 0.50	$ 20	$ 10
Misc. Costs	$ 0.30 per piece	$ 0	$ 2000
Total	$ 800	$ 2000	$ 4000

Depending on the quantity required, the total cost for the manufacturer to hedge with third-party producers is $6,800. By comparing the cost of hedging with the loss that it would incur, the manufacturer can make an informed decision when considering this strategy as one of the options by which to manage and treat the identified transactional risk in the business activity. The above mentioned approach only works in scenarios where the unserved demand over the time period of the business activity can be ascertained in crisp financial amounts. In some cases, the manufacturer will not be able to ascertain the unserved demand of the time space in crisp financial amounts. Without such an analysis, the cumulative loss due to the unserved demand of a time period cannot be determined accurately. In such cases, for risk treatment by using this option, we propose that a probabilistic approach is used to perform the comparative analysis. We discuss this approach in the next section.

9.6 Comparative Analysis to Hedge with Third-Party Producers When the Unserved Demand Is Represented in Probabilistic Amounts

When the unserved demand in the time space of the business activity has to be represented in probabilistic amounts, the steps to decide whether to consider hedging as an option for risk treatment and management are [2]:

(a) accurately determine the probability of experiencing different level/s of financial loss in the time space (Expected Loss Curve - ELC)

(b) determine the cost of hedging the production of unserved demand from third-party producers (Cost of Hedging Curve -CHC)

We discuss these steps further in the next sub-sections.

9.6.1 Ascertaining the Expected Loss Curve (ELC) of the Business Activity

We propose that the manufacturer accurately determine the probability of experiencing different level/s of financial loss in the time space by utilizing the part of the EDEC after the MAD. This is represented by the Expected Loss Curve (ELC) of the business activity. To ascertain the ELC, the manufacturer should determine the individual probability of the occurrence of different levels of unserved demand from the EDEC and then normalize them such that their cumulative probability is equal to 1. In other words, the cumulative sum of the probability of occurrence of each level of resource on the EDEC after the MAD should satisfy the condition in Equation 9.1 [2].

$$\sum_{n=1}^{z} p\,(x_n) = 1 \tag{9.1}$$

where: z represents the number of levels on the EDEC after the MAD, x_n represents each level of resources after the MAD on the EDEC, and $p(x_n)$ represents the probability of occurrence of the level of resource.

To illustrate with an example, let us consider the EDEC as shown in Figure 9.6 and the manufacturing scenario discussed in the last section. Let us assume that the MAD is $3,000. The probability mass function of the EDEC after the MAD is represented in Figure 9.9 [2].

To obtain a probabilistic curve that represents the unserved demand of the time space, the probability mass function of the curve shown in Figure 9.9 should be normalized to satisfy equation 9.1. Based on the normalized values, the ELC of the business activity represents the probability of occurrence of each level of unserved amount in the time space of the business activity. Figure 9.10 represents the ELC from the above example [2].

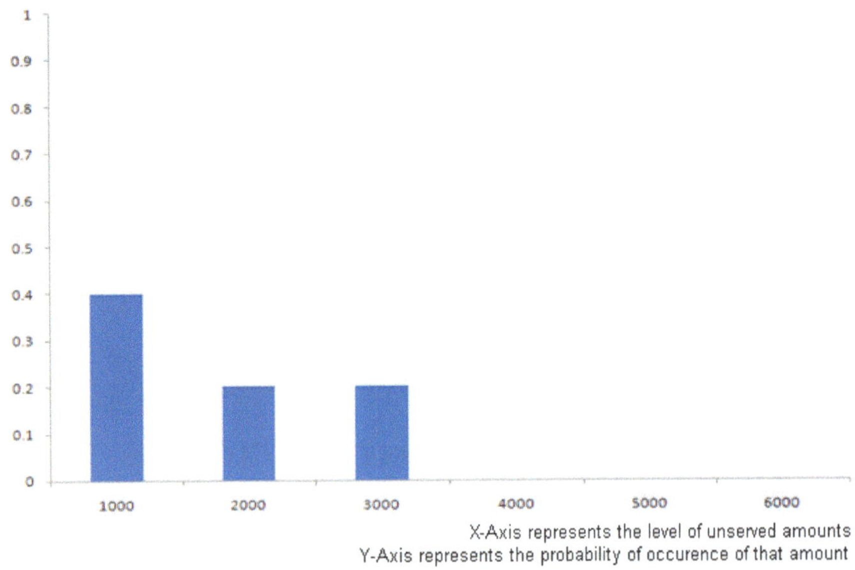

Fig. 9.9 Probability Mass Function of the EDEC after the MAD

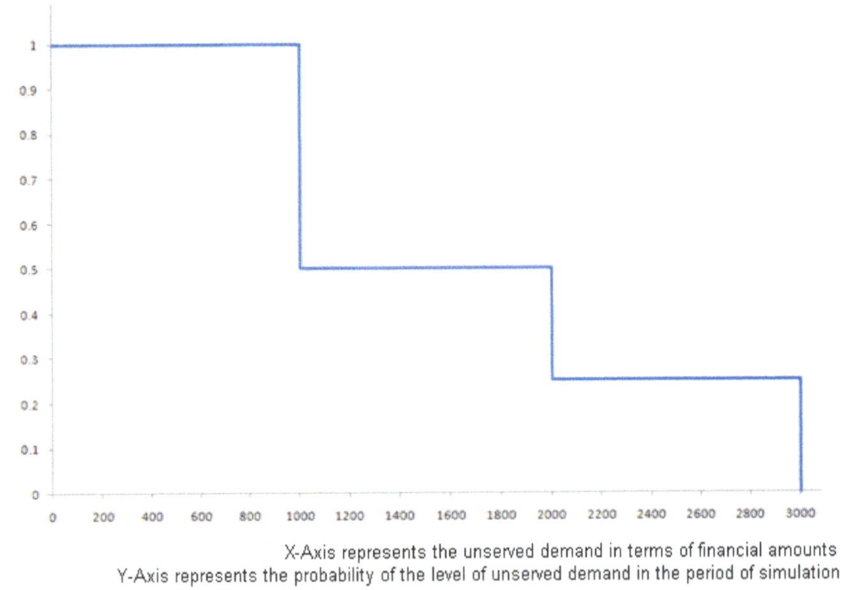

Fig. 9.10 Probabilistic curve showing the level of unserved demand in the time space

9.6.2 *Ascertaining the CHC of the Business Activity*

To perform a comparative analysis, the manufacturer should compare the ELC with the cost of alleviating this by hedging production from third-party producers, using the Cost Hedging Curve (CHC). As mentioned earlier, by analyzing the expected downtime of the production units and uncertainty in the availability of raw materials, the manufacturer will be able to determine the various product types and their quantity to which it will not be able to commit to produce in the time space and for which it needs to depend on other producers. Utilizing this, the manufacturer should determine the cost that it would incur when hedging the production of the goods with the third-party producers, in each time slot of the time space and plot the CHC. Extending the example from the previous sub-section, let us consider that the costs of hedging the unserved demand with third-party manufacturers in each time slot of the time space is as shown in Table 9.2 [2].

Table 9.2 Operational Costs to hedge with third-party manufacturers in each time slot

Manufacturing costs in / For	Time Slot T1	Time Slot T2	Time Slot T3	Time Slot T4	Time Slot T5
Product 1	200	200	100	200	100
Product 2	500	500	500	500	0
Product 3	1000	1500	1000	0	500
Total Costs	1700	2200	1600	700	600

Based on the total amount to be incurred in each time slot, the CHC is as shown in Figure 9.11 [2].

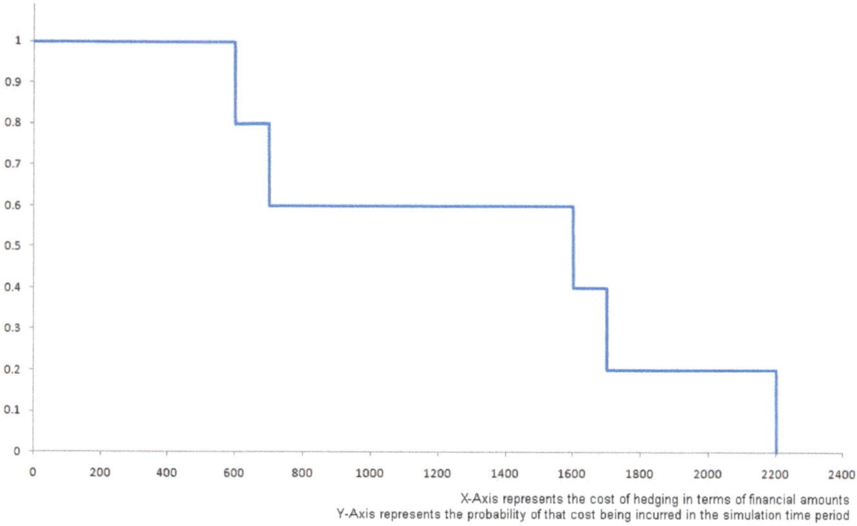

Fig. 9.11 The CHC to meet the unserved demand

In order to arrive at an informed decision, the manufacturer should compare the CHC with the ELC and decide whether to consider this method as an option for risk management. This can be done by comparing the area of the ELC with the CHC as shown in Equations 9.2 and 9.3, respectively.

$$\text{Area of Expected Loss Curve} = \int_0^n ELC \tag{9.2}$$

where: n represents the amount till the ELC extends.

$$\text{Area of Cost of Hedging Curve} = \int_0^m CHC \tag{9.3}$$

where: m represents the amount till the CHC extends.

If ELC > CHC, then the costs which the manufacturer would incur from third-party producers to meet the unserved demand is less than the loss that it would experience as a result of that unserved demand. In such cases, the manufacturer can consider hedging with third-party manufacturers as one of the options to meet the total demand for the time space. If, on the other hand, CHC > ELC, this means that the costs that the manufacturer would incur from third-party manufacturers to alleviate the loss as a result of unserved demand is more than the loss that it would actually experience due to the unserved demand. In such cases, utilizing such an option may not be a desirable approach for the manufacturer in treating and managing transactional risk and it should consider other available options to meet the unserved demand.

By utilizing the proposed approach, the manufacturer can make an informed decision about whether or not to hedge with third-party producers in order to meet the total required demand for the time space. It can then make appropriate decisions about collaborating with different third-party producers so that it can increase its production capability, enabling it to commit to the expected demand of a given time frame.

9.7 Recapitulation

As mentioned earlier, the strategies for transactional risk management are business activity-specific and hence, it is not possible to have a generalized approach for this phase that can be applied to every business activity. However, it is possible to have a generalized framework which measures the adjusted transactional risk in the business activity as a result of implementing those strategies. In this chapter, we discussed how the proposed techniques of transactional risk assessment and evaluation proposed in the earlier chapters of this book can be used for transactional risk management in a case study of just-in-time production. The strategy which we considered was to perform a comparative analysis to decide whether or not to hedge with third-party producers, in order to meet the unserved demand of the time space. The approach we discussed determines in (a) crisp amounts and (b) probabilistic terms, the cost that the manufacturer would incur in hedging the goods with the third-party producer(s) over a given time frame. The manufacturer can then compare the results of this analysis with the loss that it would experience due to the unserved demand in that time frame and then make an informed decision about whether to proceed with hedging or not. Once transactional risk is treated and managed and if the risk assessing agent decides to proceed in the business activity with the risk assessed agent, then the next phase in the process of risk analysis is transactional risk recording. We discuss this and elaborate on this phase in the next chapter.

References

1. Hussain, O.K., Dillon, T., Hussain, F., Chang, E.: Probabilistic Assessment of Loss in Revenue Generation in Demand-Driven Production. Journal of Intelligent Manufacturing (2011), doi:10.1007/s10845-011-0518-4
2. Hussain, O., Dillon, T.: Cost-Benefit Analysis to Hedge with Third-Party Producers in Demand-Driven Production. In: Xhafa, F., Barolli, L., Papajorgji, P.J. (eds.) Complex Intelligent Systems and Their Applications, vol. 41, pp. 69–81. Springer, New York (2010)
3. Simkins, S., Maier, M.: Using Just-in-Time Teaching techniques in the principles of economics course. Social Science Computer Review 22, 444–456 (2004)
4. Chen, M., Dubrawski, R., Meyn, S.P.: Management of Demand-Driven Production Systems. IEEE Transactions on Automatic Control 49, 686–698 (2004)
5. Lebovitz, R., Graban, M.: The Journey Towards Demand Driven Manufacturing. In: 2nd International Workshop on Engineering Management for Applied Technology, Texas, pp. 29–35 (2001)

Chapter 10
Risk Recording Phase: Transactional Risk Recording in Business Activities

10.1 Introduction

In the previous chapters, we proposed an approach for transactional risk identification, assessment, evaluation and management in relation to business activities. Depending on the outcome of the analysis of each phase, if the risk assessing agent initiates and completes the business association with the risk assessed agent, then, based on his experience, he should complete the risk recording phase. This phase ascertains the Actual FailureLevel (AFL) or the actual performance risk that was present in the risk assessing agent's business activity with the risk assessed agent. As explained next, in the context of business activities this phase is important as it will provide important inputs to the process of risk assessment in future business activities.

As discussed in Chapter 5, before initiating a business activity, the risk assessing agent utilizes the FailureLevel and the Failure Scale to ascertain semantically the performance risk in interacting with the risk assessed agent. The two methods by which the risk assessing agent ascertains the FailureLevel of the risk assessed agent are: utilizing his direct past business activity history; and utilizing the recommendations from other agents. In both methods, the risk assessing agent considers either the commitment level of the risk assessed agent in the particular assessment criterion/criteria or the AFL of the risk assessed agent from the risk set, that was determined according to his own or the recommending agent's past experience in interacting with the particular risk assessed agent. In other words, the risk assessing agent can determine the probability of failure in interacting with a risk assessed agent by using his direct past business activity history, only if he ascertains the actual performance risk in his past business activity with the risk assessed agent. Similarly, he can determine the probability of failure in interacting with a risk assessed agent by the using reputation-based method, only if the recommending agents which are communicating their recommendations have analyzed and summarized their experience in interacting with that risk assessed agent, based on which they form the risk set. Subsequently, it is important that the risk assessing agent, after interacting with a risk assessed agent, ascertains the actual performance risk of that agent in their business activity, so that he can either utilize it in his future business activity with that risk

O.K. Hussain et al.: Risk Assessment & Management in the Networked Economy, SCI 412, pp. 263–281.
springerlink.com

assessed agent, or act as the recommending agent to communicate this to a querying agent.

In this chapter, we present an approach by which the risk assessing agent can ascertain and record the level of transactional risk present in the business association with the risk assessed agent. As was the case in the assessment of transactional risk in the business activity, the process of its recording, too, should be strictly according to the specific characteristics, expectations and time period of the business activity. In the next sections, we present the approach by which these characteristics of transactional risk are identified and the level of actual transactional risk determined and recorded.

10.2 Stages in the Transactional Risk Recording Phase

The two stages in the transactional risk recording phase are as follows:

10.2.1 Determine the Actual FailureLevel (AFL) of the Risk Assessed Agent

The aim of this step is to ascertain the level of performance risk or Actual FailureLevel (AFL) in the risk assessing agent's business activity with the risk assessed agent. The advantages of this step are as follows:

- If the risk assessing agent wants to interact again with the particular risk assessed agent in the same time slot and with the same characteristics as that of his just concluded business activity, he can utilize the determined level of performance risk (AFL) in his previous business activity, as the performance risk of the risk assessed agent in the particular time slot of his future business activity.
- Similarly, if the risk assessing agent wants to interact with the particular risk assessed agent with the same characteristics as the previous business activity, and if the time of his previous business activity is within the time slots which come in the pre-interaction start time phase of his future business activity, then while determining the FailureLevel (PFL) of the risk assessed agent in that time slot(s), the risk assessing agent can consider his past business activity history with that agent rather than determining it from the recommending agents.
- If the expectations of the future business activity of the risk assessing agent with the risk assessed agent partially meets the criteria of his previous business activity, he can consider the commitment level of the risk assessed agent in those partial assessment criterion/criteria, according to his past business activity experience, and solicit recommendations for the other assessment criterion/criteria from other agents to determine the FailureLevel (PFL or FFL) of the risk assessed agent.
- Based on the determined AFL value, the risk assessing agent can act as a recommending agent and communicate his recommendation about the risk

assessed agent when solicited by other agents in the context, time and assessment criteria of his business activity.

10.2.2 Adjust the Credibility Value (RCV) of the Recommending Agents

This step involves adjusting or updating the credibility of the recommending agents after the business activity, if the recommending agent took any. The advantage of this step is:

- As discussed in Chapter 5, the risk assessing agent, while determining the FailureLevel of the risk assessed agent (PFL and FFL) by the reputation-based method, considers and evaluates recommendations from those agents whose credibility is either trustworthy or unknown, and omits recommendations from those agents whose credibility is untrustworthy. Subsequently, after the business activity the risk assessing agent should adjust the credibility of the recommending agents from which he took recommendations, while determining the FailureLevel of the risk assessed agent. Doing so enables the risk assessing agent to maintain an updated status of the recommending agents in the particular context and allows him to consider future recommendations from them, according to their credibility.

An important point to note here is that the credibility adjustment of a recommending agent should be strictly according to assessment criterion/criteria against which the risk assessing agent considered his recommendation, and not according to the total assessment criteria of his business activity with the risk assessed agent.

In the next sections of this chapter, we propose an approach by which the risk assessing agent can undertake the above mentioned steps in the risk recording phase. In Sections 10.3 – 10.7, we propose and explain our approach by which the risk assessing agent summarizes his experience and determines the AFL of the risk assessed agent in the business activity. In Sections 10.8 – 10.9, we propose our approach by which the risk assessing agent, based on his experience in the business activity, adjusts the credibility of the recommending agents strictly according to the assessment criteria from which he took recommendations.

10.3 Definition of Key Terms Used While Recording Transactional Risk

10.3.1 Expectations

As discussed in Chapter 5, the *Expectations* of the business activity represent how the risk assessing agent and the risk assessed agent should commit to fulfill the assessment criteria in the business activity. In other words, the expectations represent the *promised commitment* or *promised behavior* formed between the two agents before initiating their business activity.

10.3.2 Promised Behaviour (ProCom $_{\text{Business Activity}}$)

Promised Behavior (ProCom$_{\text{Business Activity}}$) quantifies the behavior to which the risk assessed agent promised to commit in the business activity. In other words, this is the promised commitment or the quantification of the expectations in the business activity.

10.3.3 Actual Behavior (Assess $_{\text{Business Activity}}$)

Actual Behavior (Assess $_{\text{Business Activity}}$) refers to the behavior or commitment which the risk assessed agent showed in the business activity.

10.3.4 Non-commitment (Failure $_{\text{Business Activity}}$)

Non-commitment (Failure $_{\text{Business Activity}}$) refers to the degree to which the risk assessing agent's expectations are not met by the risk assessed agent in the business activity. Calculating this will assist the risk assessing agent to ascertain the degree to which the risk assessed agent has not met the expectations of the business activity.

10.3.5 Recommending Agent's Credibility Value (RCV)

As mentioned in Chapter 5, *Recommending Agent's Credibility Value* (RCV) represents the credibility of an agent while giving recommendations in a certain context.

10.4 Process of Ascertaining the Actual FailureLevel (AFL) in a Business Activity

The method by which the risk assessing agent summarizes and determines the AFL of the risk assessed agent in the business activity after interacting with him is as follows[1, 2]:

- quantifying the expectations of the business activity formed between the two agents,
- assessing the level of non-commitment in the risk assessed agent's actual behavior compared to the expectations of the business activity.

To do this, the risk assessing agent should first determine the level of commitment that the risk assessed agent showed in the business activity by analyzing his actual behavior. The actual behavior depicts the extent to which the risk assessed agent fulfilled or committed to the expectations in the business activity. If the level of commitment (i.e. the actual behavior) is compared with the expectations (i.e. promised commitment or promised behavior), then the non-committed behavior in the business activity can be determined. The degree of non-committed behavior is used to determine the AFL of the business activity on the Failure Scale. The greater the non-committed behavior, the greater is the actual level of failure in a business activity and vice versa.

In the next section, we define the metrics by which the risk assessing agent ascertains the level of commitment or the actual behavior of the risk assessed agent in each assessment criterion from the expectations of his business activity.

10.5 Metrics for Ascertaining the Actual Behavior by the Risk Assessed Agent in a Business Activity

In this section, we propose the metrics by which the risk assessing agent quantifies the actual behavior or the level of commitment of the risk assessed agent after interacting with him.

10.5.1 Assessment of Total Commitment in a Business Activity (Assess $_{Business\ Activity}$)

We quantify the level of commitment or the actual behaviour by the risk assessed agent in the business activity by the metric Assess $_{\text{Business Activity}}$. As discussed in Chapter 5, the expectations of a business activity are composed of the different assessment criteria, along with the quality metrics that the risk assessing agent wants to achieve, in the business activity. Subsequently, the risk assessing agent, in order to ascertain the actual behaviour or the level of commitment by the risk assessed agent in a business activity, (Assess $_{\text{Business Activity}}$) has to [2]:

- evaluate the level of commitment by the risk assessed agent in each assessment criterion of the business activity;
- add the evaluations of all the assessment criteria to get the total commitment or the actual behaviour by the risk assessed agent in the business activity (Assess$_{\text{Business Activity}}$).

The actual behaviour or the level of commitment of the risk assessed agent in the business activity (Asses $_{\text{Business Activity}}$) is determined by the evaluation of individual commitment in each assessment criterion. Hence:

$$\text{Assess}_{\text{Business Activity}} = \sum_{i=1}^{n} \text{Eval}_{\text{Assessment Criterion I}} \tag{10.1}$$

where:
n represents the number of assessment criteria in the business activity.

10.5.2 Evaluation of an Assessment Criterion in an Business Activity ($Eval_{Assessment\ Criterion}$)

$Eval_{Assessment\ Criterion}$ is defined as the metric which evaluates the degree of fulfilment of an assessment criterion in the actual behaviour of the risk assessed agent with respect to the promised behaviour of the business activity. In other words, the metric $Eval_{Assessment\ Criterion}$ is used to determine whether or not the particular assessment criterion has been fulfilled, in accordance with the promised behaviour by the risk assessed agent in the actual behaviour. To evaluate the degree to which an assessment criterion was fulfilled with respect to the expectations, two levels are defined for the metric $Eval_{Assessment\ Criterion}$ as shown in Table 10.1 [2]. Each of these two levels corresponds to a different level of fulfilment of the assessment criterion. A numerical value is assigned to each level, and the value which corresponds to the level to which the assessment criterion was fulfilled by the risk assessed agent is taken into account while evaluating that criterion.

Table 10.1 Levels for the metric $Eval_{Assessment\ Criterion}$

$Eval_{Assessment\ Criterion}$ Value	Semantics of the Value
0	The risk assessed agent did not fulfil the assessment criterion according to the promised behaviour.
1	The assessment criterion was fulfilled exactly according to the promised behaviour by the risk assessed agent.

While evaluating the fulfilment of each assessment criterion in order to determine the AFL of the risk assessed agent, it is also important for the risk assessing agent to consider some other factors. We will explain these in the next sub-section and define the metrics to quantify them so that they can be measured and considered while determining the AFL of the risk assessed agent.

10.5.3 Accuracy of Assessment Criterion Communication in an Business Activity ($Accu_{Assessment\ Criterion}$)

After evaluating an assessment criterion in relation to the expectations, the risk assessing agent can consider it to ascertain the AFL of the risk assessed agent, only if that assessment criterion has been evaluated against the quality metrics which has been communicated to the risk assessed agent in *clear terms beforehand*. For example, as shown in Figure 4.5, it is not fair to the risk assessed agent if he is evaluated against a criterion which was not communicated to him clearly. So, it is important that the risk assessing agent communicates the quality metric/s for each assessment criterion clearly to the risk assessed agent beforehand, in the expectations. So after the business activity has concluded, it can be evaluated and subsequently, the risk assessed agent can be assigned a FailureLevel value that he actually deserves, based on his commitment to the criteria.

Hence, the Accuracy of Assessment Criterion Communication metric ($Accu_{Assessment\ Criterion}$) is defined as the metric which is used to determine whether or not the quality metric which the risk assessing agent utilized to evaluate an assessment criterion was communicated in clear terms to the risk assessed agent beforehand, in the expectations. In order to quantify the accuracy by which the quality metric of an assessment criterion was communicated to the risk assessing agent beforehand, two levels are defined for the metric $Accu_{Assessment\ Criterion}$, as shown in Table 10.2 [2]. Each numerical value corresponds to a level of accuracy with which the utilized quality metric was used to evaluate the assessment criterion. This will be taken into consideration while evaluating that assessment criterion.

Table 10.2 Levels for the metric $Accu_{Assessment\ Criterion}$

$Accu_{Assessment\ Criterion}$ Value	Semantics of the Value
0	The quality metric against which the assessment criterion has been evaluated or judged has not been communicated to the risk assessed agent beforehand in clear terms.
1	The quality metric against which the risk assessment criterion has been evaluated or judged has been communicated to the risk assessed agent beforehand in clear terms.

10.5.4 Significance of an Assessment Criterion ($Sig_{Assessment\ Criterion}$)

Another important factor to consider while assessing the actual behaviour or the level of commitment by the risk assessed agent is the significance of each assessment criterion. As discussed in Section 4.3, all the assessment criteria of the expectations will not be of equal importance. The risk assessing agent may regard some assessment criteria as being crucial to his outcomes, hence if these are not fulfilled, this may result in a high level of failure and subsequently, represents a high performance risk in relation to the risk assessed agent. On the other hand, the risk assessing agent may consider some assessment criteria to have a smaller impact on the outcome of the business activity. In order to take such cases into consideration, the risk assessing agent should ascertain the FailureLevel (PFL and FFL) of the risk assessed agent, according to the significance of each assessment criterion, from his expectations.

Similarly, while determining the Actual FailureLevel (AFL) of the risk assessed agent, the risk assessing agent should consider each assessment criteria, according to its significance. The metric $Sig_{Assessment\ Criterion}$ is defined in order to take into consideration the significance of the particular assessment criterion while ascertaining the level of commitment or the actual behaviour of the risk assessed agent in the business activity. When determining the AFL of the risk assessed agent, the significance value of each assessment criterion will be the same as what assigned to it by the risk assessing agent, before commencing his business activity with the risk assessing agent. These significance levels for each assessment criterion are defined by using the approach proposed in Section 4.4.

10.5.5 Quantifying the Actual Behavior in the Business Activity ($Assess_{Business\ Activity}$)

Once the risk assessing agent assigns the corresponding value of each metric for each assessment criteria of the expectation, then the actual behaviour or the level of commitment of the risk assessed agent can be determined by considering each assessment criteria in relation to:

- the degree to which the assessment criterion has been met,
- the accuracy of communication of the quality metric of the assessment criterion, and
- the significance of the assessment criterion.

Hence, the total commitment of the risk assessed agent in the business activity is expressed as:

$$\mathbf{Assess}_{\mathbf{Business\ Activity}} = \sum_{i=1}^{n} (\text{Eval}_{\text{Assessment Criterion i}} * \text{Accu}_{\text{Assessment Criterion i}} * \text{Sig}_{\text{Assessment Criterion i}}) \qquad (10.2)$$

where:
n represents the number of assessment criteria in the business activity.

Once the risk assessing agent determines the actual behaviour of the risk assessed agent in the business activity, he has to then compare it with the expectations or the promised behaviour of his business activity to determine the level of non-commitment, and subsequently the AFL of the risk assessed agent. In the next section, we propose an approach by which the risk assessing agent ascertains the level of non-committed behaviour and subsequently the AFL of the risk assessed agent.

10.6 Ascertaining the Level of Non-commitment and the AFL in the Business Activity

In the previous section, we defined a methodology by which the risk assessing agent quantifies numerically the actual behaviour or the level of commitment by the risk assessed agent in the business activity. To determine the degree of non-commitment and the AFL in the business activity, the risk assessing agent should also quantify numerically the promised behaviour in the business activity. We discuss this further in the next sub-section.

10.6.1 Quantifying the Promised Commitment in the Business Activity (ProCom $_{Business\ Activity}$)

As discussed earlier, the expectations of the business activity are defined as the promised behaviour or the promised commitment to which the risk assessed agent is expected to commit in the business activity. We represent the expectations or the promised behaviour in the business activity by the metric $\text{ProCom}_{\text{Business Activity}}$. The promised behaviour or the promised commitment by the risk assessed agent in the business activity can be quantified by substituting a value of 1 for the metric $\text{Eval}_{\text{Assessment Criterion}}$ in equation 10.2, for each assessment criteria of the expectations. This value for the metric suggests that the assessment criteria of the business activity have been committed according to the promised behaviour, subsequently the summation of which quantifies the promised behaviour in the business activity as the result of the fulfilment of each assessment criteria according to the expectations. So, the promised behaviour or the promised commitment in the business activity can be quantified as:

$$\textbf{ProCom}_{\textbf{Business Activity}} = \sum_{i=1}^{n} (1 * \text{Accu}_{\text{Assessment Criterion i}} * \text{Sig}_{\text{Assessment Criterion i}}) \qquad (10.3)$$

where:
n represents the number of assessment criteria in the expectations.

10.6.2 Ascertaining the Non-commitment in the Business Activity (Failure $_{Business\ Activity}$)

We define the level of non-commitment in the business activity by the metric $\text{Failure}_{\text{Business Activity}}$. This metric expresses the actual level of failure in the business activity due to the degree of non-commitment by the risk assessed agent in his actual behaviour versus the promised behaviour. The level of non-commitment in the business activity is found by the difference between the promised behaviour ($\text{ProCom}_{\text{Business Activity}}$) and the actual behaviour ($\text{Assess}_{\text{Business Activity}}$). Hence, $\text{Failure}_{\text{Business Activity}}$ is quantified as:

$$\textbf{Failure}_{\textbf{Business Activity}} = \frac{\Pr oCom_{Bu\ \sin\ essActivity} - Assess_{Bu\ \sin\ essActivity}}{\Pr oCom_{Bu\ \sin\ essActivity}} * 100 \qquad (10.4)$$

In other words, the metric $\text{Failure}_{\text{Business Activity}}$ shows the degree of failure or the degree of non-commitment in the actual behaviour of the risk assessed agent when compared to the promised behaviour. Note that $\text{ProCom}_{\text{Business Activity}}$ in the denominator would be zero if either $\text{Accu}_{\text{Assessment Criterion i}}$ or $\text{Sig}_{\text{Assessment Criterion i}}$ are zero for all i. However, this would not correspond to a sensible promised commitment.

10.6.3 Ascertaining the Level of Non-commitment in the Business Activity ($AFL_{\text{Business Activity}}$)

To determine the AFL of the risk assessed agent on the Failure Scale, the risk assessing agent has to map the $\text{Failure}_{\text{Business Activity}}$ to the Failure Scale which ranges from (-1, 5), as shown in Figure 5.5. However, the level of -1 on the Failure Scale shows that the level of failure in the business activity is unknown. This level cannot be assigned by the risk assessing agent to the risk assessed agent after interacting with him. After the business activity, the risk assessing agent can assign the risk assessed agent with a FailureLevel value between (0, 5) on the Failure Scale. The FailureLevel which corresponds to the degree of failure present in the business activity, when mapped against the Failure Scale, is taken as the Actual FailureLevel (AFL) of the risk assessed agent. Hence, the AFL of the risk assessed agent in the business activity is determined as [2]:

$$\textbf{Actual FailureLevel (AFL)} = \text{LEVEL}\ (\text{Failure}_{\text{Business Activity}}) \tag{10.5}$$

Equation 10.5 can also be written as:

$$\textbf{Actual FailureLevel(AFL)} =$$
$$\text{LEVEL}\ \left(\frac{\Pr oCom_{Bu\sin essActivity} - Assess_{Bu\sin essActivity}}{\Pr oCom_{Bu\sin essActivity}} * 100\right)$$

The proposed concept and its significance will be explained more fully in the next section with an example.

10.7 Example of Determining the AFL in the Business Activity

In this section, we explain the process whereby the risk assessing agent determines the AFL of the risk assessed agent in an business activity by using the above metrics. To proceed further, let us consider the business activity scenario discussed in Chapter 4, where the risk assessing agent ‘A’ wanting to interact with a logistics company in the context of ‘Transporting Goods’ (termed as context ‘C’) from 26/03/2010 to 02/04/2010. The expectations formed in the business activity are outlined in Section 4.2.2.1.Let us consider that the risk assessing agent ‘A’ assigns a significance value of 0.3, 0.2, 0.1 and 0.4 to assessment criterion C1, C2, C3 and C4 respectively. Furthermore, in order to explain the effectiveness of the model better, let us suppose that risk assessing agent ‘A’, by his assessment criterion C4: *‘The logistics company should provide him with a track and trace facility’,* meant that the facility should be provided throughout the period of his business activity. However, as seen from the expectations, the particular quality metric for the assessment criterion was not communicated in clear terms.

The possible agents with which risk assessing agent 'A' can interact are logistics companies 'B', 'D' and 'E'. Let us consider that, based on analysing the transactional risk in interacting with each agent, risk assessing agent 'A' chooses agent 'E' with which to interact. After his business activity, agent 'A' should determine the AFL in interacting with agent 'E'. Let us suppose the actual behaviour of risk assessed agent 'E' in the business activity was as follows:

- delivered the goods to their destination undamaged (C1);
- delivered the goods by 5 pm on the specified day (C2);
- unpacked the goods at the destination (C3);
- provided track and trace facility only for the period of transit to the destination (C4).

Risk assessing agent 'A' has to assess the level of fulfilment or commitment in the actual behaviour of agent 'E' with respect to the expectations or promised behaviour. The value of $Eval_{Assessment\ Criterion}$ can be determined according to its metric as follows:

- For the first assessment criterion, risk assessed agent 'E' delivered the goods undamaged to their destination and, hence fulfilled the assessment criterion, according to the expectations. So the value of $Eval_{C1}$ is 1.
- For the second assessment criterion, risk assessed agent 'E' delivered the goods by 5 pm on the specified day as opposed to 8 am as promised in the expectations. The value of $Eval_{C2}$ is 0.
- For the third assessment criterion, risk assessed agent 'E' unpacked the goods at the destination. Hence, the value of $Eval_{C3}$ is 1.
- For the fourth assessment criterion, agent 'E' did not provide a track and trace facility throughout the business activity. Therefore, the risk assessing agent evaluates $Eval_{C4}$ as 0.

After evaluating the fulfilment of each assessment criterion, agent 'A' should determine the accuracy with which the quality metric against which each assessment criterion has been evaluated, was communicated in the expectation. For this, the metric $Accu_{Assessment\ Criterion}$ will be used:

- Assessment Criterion C1 was communicated clearly - the value of $Accu_{C1}$ is 1.
- C2 was communicated clearly - the value of $Accu_{C2}$ is 1.
- C3 was communicated clearly - the value of $Accu_{C3}$ is 1.
- C4 was not communicated clearly - the value of $Accu_{C4}$ is 0.

After assigning the relative numerical values of each metric to all the assessment criteria, the actual behaviour of risk assessed agent 'E' in the business activity ($Assess_{Business\ Activity}$) quantified by using equation 10.2 is:

$$\mathbf{Assess_{Business\ Activity} = 0.4}$$

Similarly, the promised behaviour of agent 'E', according to the expectations formed about the business activity quantified by equation 10.3, is:

$$\mathbf{ProCom_{Business\ Activity} = 0.7}$$

The degree of non-commitment in the actual behaviour of the risk assessed agent as compared to the promised behaviour is determined by using equation 10.4 as:

Failure $_{\text{Business Activity}}$ = 42.85 %

Using equation 10.5, we map the value of the metric Failure $_{\text{Business Activity}}$ on the Failure Scale to determine the AFL of the risk assessed agent in the business activity:

Actual FailureLevel (AFL) = 3

Based on his assessment, risk assessing agent 'A' can form the risk set which summarizes his business activity experience with risk assessed agent 'E'. The risk set, based on the assessment of the above business activity, is:

{Agent 'A', Agent 'E', Context 'C', 3, ((C1, 1), (C2, 0), (C3, 1)), 10000, 26/03/2010, 02/04/2010}

From the above example, it can be seen that:

- the quality metric by which assessment criterion 'C4' has been evaluated was not communicated to risk assessed agent 'E' accurately in the expectations. Subsequently, this assessment criterion was not included while determining the AFL of the risk assessed agent, by not taking it into consideration while ascertaining the $\text{ProCom}_{\text{Business Activity}}$ of the business activity.
- As can be seen from the risk set, assessment criterion 'C4' is not included in the risk set as risk assessing agent 'A' failed to communicate it clearly to risk assessed agent 'E' and hence, while communicating a recommendation for that agent, agent 'A' does not include it in the risk set.

Suppose that when an agent, say agent 'F' solicits recommendations about agent 'E' in the same context, then agent 'A' can reply with the risk set that he formed in his business activity, provided that the time and assessment criteria for which the recommendation is solicited matches partially or completely those of agent 'A' in his business activity. Upon receiving the recommendation from agent 'A' in the form of a risk set, agent 'F' appends agent 'A's RCV so that he can classify the recommendation, according to his credibility. Subsequently, it is imperative that after his business activity, agent 'F' should adjust the credibility of recommending agent 'A' from whom he took recommendations, so as to maintain his accurate credibility or RCV. In the next section, we propose a methodology by which the risk assessing agent of a business activity adjusts the credibility of the recommending agents from which he took recommendations, after interacting with a risk assessed agent.

10.8 Adjusting the Credibility of the Recommending Agents

As discussed in Chapter 5, it may be possible that the risk assessing agent may consider recommendations in partial assessment criteria of his expectations when ascertaining the PFL or FFL of a risk assessed agent. If this is the case, then after the business activity, the credibility adjustments of the recommending agents should be done only by considering those assessment criterion or criteria from which the risk assessing agent took his recommendation, and not by considering the total assessment criteria of his expectations. In the next sub-sections, we discuss the process of RCV adjustment of a recommending agent.

10.8.1 Ascertaining the Adjustment to Be Made to the RCV of the Recommending Agent

Adjusting the RCV of a recommending agent 'R' based on his recommendation in an assessment criterion 'Cn' can be done by determining the difference between (a) and (b) below[3]:

(a) the assessment criterion commitment that the risk assessing agent found after interacting with the risk assessed agent, and
(b) what agent 'R' recommended to the risk assessing agent for that particular assessment criterion before his interaction.

If there is a discrepancy between the commitment level of an assessment criterion that the risk assessing agent found after his interaction and the commitment level communicated by the recommending agent prior to interaction, we propose that:

(a) the magnitude of adjustment in the credibility of the recommending agent should be more if he was a *trustworthy* recommending agent, compared to if he was an *unknown* recommending agent, at the time of communicating the recommendation. This is consistent with the methodology mentioned in Section 5.9, where the risk assessing agent gives more weight to recommendations from agents who are trustworthy, compared to ones who are unknown, while determining either the PFL or FFL of the risk assessed agent. Hence, if there is any deviation in the recommendation, then after his interaction, we propose that the credibility adjustment of the recommending agent should also be calculated accordingly. This is doneby the weight 'C_R' in equation 10.6.
(b) as mentioned in Section 5.7, while ascertaining the PFL of a risk assessed agent, the risk assessing agent gives more weight to those recommendations which are in the time slots near to the time spot of his business activity. Therefore, if he finds any discrepancies in those recommendations after his interaction, then the credibility adjustment should be calculated accordingly. This is done by the variable 'w' in equation 10.6.
(c) the adjustment to the recommending agent's credibility should be made according to the significance of a particular assessment criterion.

The adjustment to be made to the recommending agent 'R's' credibility, according to the recommendation that he gave to the risk assessing agent 'A' in assessment criterion 'Cn' and in time slot 't-z' can be determined by:

$$(A^{R}_{Cnt-z})^{1}_{-1} = (C_R * (w * (\mathrm{Com}^{Cn}{}_{A} - \mathrm{Com}^{Cn}{}_{R}))) \tag{10.6}$$

where:
$\mathrm{Com}^{Cn}{}_{A}$ represents the commitment level of the risk assessed agent, determined by the risk assessing agent 'A' in assessment criterion 'Cn', according to his experience after the interaction,
$\mathrm{Com}^{Cn}{}_{R}$ represents the commitment level recommended by the recommending agent 'R' to the risk assessing agent 'RA' for the risk assessed agent in assessment criterion 'Cn' before the interaction,
'w' is the weight which the risk assessing agent had applied to consider the status of the risk assessed agent in time slot 't-z' in equation 5.4,
'C_R' is the variable which is applied to the credibility adjustment of the recommending agent, depending on whether he was a trustworthy or unknown recommending agent at the time of communicating the recommendation. C_R is represented by two variables, α and βwhose values are the same as the risk assessing agent considers in equation 5.4.

The deviation in the recommendation, when weighted with the credibility of agent 'R' at the time of the recommendation and the time slot of his recommendation gives the accurate measure of adjustment that is to be made to the RCV of the recommending agent. It is possible that the risk assessing agent in a time slot may take recommendations from a recommending agent in more than one assessment criterion. In this case, after the interaction, the risk assessing agent, by using equation 10.6, should determine the degree of adjustment to be made to the risk assessing agent's credibility, according to the recommendation which he gave for each assessment criterion in time slot 't-z'.

10.8.2 Adjusting the RCV of the Recommending Agent

Once the adjustment to be made to a recommending agent 'R's credibility has been determined, the risk assessing agent should then weigh it against the significance of the assessment criterion/criteria, before making the final adjustment to his RCV. As discussed in Section 4.3, all the assessment criteria of the business activity will not be of equal importance or significance. Some assessment criterion might be considered as very important for the business activity. So, the adjustment of the recommending agent's credibility, according to the recommendation that he gave in an assessment criterion, should also be calculated according to its significance.

For example, let us suppose that recommending agent 'R's recommendation in assessment criterion 'Cn' played a significant part in risk assessing agent 'A' forming a business association with risk assessed agent 'B'. Further, let us consider that agent 'A' considered assessment criterion 'Cn' as significantly important and after the business activity,he found that there is a deviation in what agent 'R' recommended and to what agent 'B' committed, thereby resulting in a negative outcome. Similarly, the RCV adjustment of agent 'R' should be done accordingly so that agent 'A' considers any future recommendations from him, according to his past experience. The adjustmentto be made to recommending agent's 'R' credibility, according to his recommendation in time slot 't-z' and according to the significance of the assessment criterion/criteria is as follows[3]:

$$A^{R}_{t-z} = \left(\sum_{i=1}^{n} S_{Cn} * A^{R}_{Cnt-z}\right) \tag{10.7}$$

where:

A^{R}_{t-z} denotes the adjustment to be made to the credibility of recommending agent 'R', according to his recommendation in time slot 't-z',

n represents the number of assessment criteria in which the risk assessing agent took recommendations from the recommending agents in time slot 't-z',

S_{Cn}represents the significance of the assessment criterion 'Cn' in which the risk assessing agent took the recommendation from the recommending agent,

A^{R}_{Cnt-z} represents the adjustment to be made to recommending agent's 'R' credibility, according to the recommendation that it gave to risk assessing agent 'RA' in assessment criterion 'Cn' and in time slot 't-z'.

Finally, the risk assessing agent can make the adjustment to the recommending agent's credibility, according to the recommendations that he gave in time slot 't-z' by using the equation [3]:

$$RCV_{\text{NEW R}} = RCV_{\text{OLD R}} \oplus A^{R}_{t-z} \tag{10.8}$$

where:

$\oplus$ is the adjustment operator.

$RCV_{\text{NEW R}}$ will now become the credibility of the recommending agent in the particular context as adjusted by the risk assessing agent, according to his recommendation. If the risk assessing agent considers the recommendation from a recommending agent in more than one time slot of its interaction, then, by using equation 10.6 – 10.8, he should ascertain the final adjustment to be made to the recommending agent's credibility, according to his recommendation in each time slot. The proposed concept will become clear after we explain it using an example in the next section.

10.9 Example of Adjusting the RCV of the Recommending Agents

Continuing our discussion from Section 10.7, let us consider that risk assessing agent 'A', in the pre-interaction start time slot 't-2' of the time space, determines the FailureLevel (PFL) of agent 'E' in assessment criteria C1 and C2 by using recommendations from other agents and in assessment criteria C3 and C4 by using his own past-interaction history. Let us consider that the recommendations which risk assessing agent 'A' receives in time slot 't-2' are:

Recommendation from agent 'I':
{Agent 'I', Agent 'E', C, 3, ((C1, 0) (C9, 0)), $8000, 18/03/2010, 25/03/2010, UNKNOWN}
Recommendation from agent 'H':
{Agent 'H', Agent 'E', C, 5, ((C2, 1) (C8, 1) (C10, 1)), $9000, 15/03/2010, 20/03/2010, 0}
Recommendation from agent 'J':
{Agent 'J', Agent 'E', C, 5, ((C1, 1) (C3, 1)), $200, 17/03/2010, 24/03/2010, -1}

Risk assessing agent 'A' appends the credibility value of each recommending agent to his risk set to evaluate his recommendation accordingly. It can be seen that the recommendation from agents 'H' and 'J' are known recommendations which are trustworthy, whereas the recommendation from agent 'I' is unknown. The level of commitment that risk assessing agent 'A' determined for risk assessed agent 'E' in the assessment criteria C1, C2 and C3 of his interaction are 1, 0, and 1 respectively (from section 10.7). In order to adjust the credibility of the recommending agents 'I', 'H' and 'J', according to the recommendation that they gave, risk assessing agent 'A' has to first determine the deviation in the assessment criterion commitment level that he found and what the recommending agents recommended for that assessment criterion.

10.9.1 Adjusting the RCV of Agent 'I'

Determining and representing the deviation in the commitment level for assessment criterion C1 between what risk assessing agent 'A' found after his interaction and what recommending agent 'I' recommended is shown in Table 10.3:

Table 10.3 Determining deviation in recommendations of agent 'I'

Assessment Criterion C1	Commitment Level
Commitment level determined by the risk assessing agent	1
Commitment level recommended by the recommending agent	0

As can be seen from the risk set, agent 'I' is an unknown recommending agent and his recommendation is in a time slot to which the risk assessing agent gives more weight, in order to consider the fresh status of the risk assessed agent. Let us assume that risk assessing agent 'A', while determining the FailureLevel (PFL) of the risk assessed agent, gives a weight of 0.6 and 0.4 to recommendations from trustworthy and unknown recommending agents respectively (α and β in equation 5.4). We utilize equation 10.6 to determine the adjustment to be made to the credibility of recommending agent 'I', according to the recommendation he gave in assessment criterion 'C1' in time slot 't-2':

$$A^{I}_{C1t-2} = 0.4$$

Let us consider that the significance of assessment criterion C1 is 0.328. We determine the adjustment to be made to recommending agent's 'I' credibility, according to the significance of the assessment criterion 'C1' in time slot 't-2', by using equation 10.7:

$$A^{I}_{t-2} = 0.133$$

Finally, we use equation 10.8 to adjust the credibility of the recommending agent 'I':

$$RCV_{NEW\,I} = 0.133$$

10.9.2 Adjusting RCV of Agent 'H'

Determining and representing the deviation in the commitment level for assessment criterion 'C2' between what risk assessing agent 'A' found after his interaction and what recommending agent 'H' recommended is shown in Table 10.4:

Table 10.4 Determining deviation in the recommendation of agent 'H'

Assessment Criterion C2	Commitment Level
Commitment level determined by the risk assessing agent	0
Commitment level recommended by the recommending agent	1

We utilize equation 10.6 to determine the adjustment to be made to the credibility of recommending agent 'H', according to the recommendation he gave in assessment criterion 'C2' in time slot 't-2':

$$A^{H}_{C2t-2} = -\ 0.6$$

Considering the significance value of assessment criterion C2 is 0.1867, we useequation 10.7 to determine the adjustment to be made to recommending agent's 'H' credibility:

$$A^{H}_{t-2} = -\ 0.112$$

Finally, we use equation 10.8 to adjust the credibility of recommending agent 'H', according to his recommendation in time slot 't-2':

$$RCV_{NEW\,H} = 0 \oplus (-0.112) = -0.112$$

After the adjustment, the credibility (RCV) of recommending agents 'I' and 'H' becomes 0.133 and -0.2, respectively. There is no change in the credibility of recommending agent 'J' as there is no deviation in the credibility recommended by that agent with what was found by risk assessing agent 'A' in the assessment criterion 'C1'. As can be seen in the examples,

- the degree and value of adjustment (+ ve or – ve) in the credibility value of a recommending agent is calculated accordingly, if the recommending agent recommended a lower and higher value respectively for an assessment criterion which is found to be incorrect by the risk assessing agent after the interaction;
- the credibility value of an agent which approaches the extremes (-1 and 1) will result in the recommending agent being classified as untrustworthy.
- based on the adjustments, recommending agent 'A' knows the accurate RCV of the recommending agents, and he can classify their recommendations according to their credibility when he considers them again in the near future.

10.10 Recapitulation

In this chapter, we discussed the importance of the transactional risk recording phase and proposed approaches by which the steps of this phase can be undertaken. Specifically, we proposed a methodology by which the risk assessing agent summarizes his experience and determines the Actual FailureLevel (AFL) that was present in interacting with a risk assessed agent. Furthermore, we also proposed a methodology by which the risk assessing agent, after summarizing his experience in interacting with a risk assessed agent, adjusts the credibility of the recommending agents, from whom he took recommendations while making an informed decision about his future course in the business activity with the risk assessed agent.

References

1. Hussain, O.K., Chang, E., Hussain, F.K., Dillon, T.S.: A Methodology for Determining Riskiness in Peer-to-Peer Communications. In: IEEE (ed.) Proceedings of the IEEE International Conference on Industrial Informatics (INDIN 2005), vol. 1, pp. 655–666. IEEE, Perth (2005)

2. Hussain, O.K., Chang, E., Hussain, F.K., Dillon, T.S.: A methodology for risk measurement in e-transactions. Computer Systems Science and Engineering 21, 17–31 (2006)
3. Hussain, O.K., Chang, E., Hussain, F.K., Dillon, T.S.: A Methodology for Determining the Creditability of Recommending Agents. In: Gabrys, B., Howlett, R.J., Jain, L.C. (eds.) KES 2006. LNCS (LNAI), vol. 4253, pp. 1119–1127. Springer, Heidelberg (2006)

Chapter 11
New Research Directions

11.1 Introduction

In the literature, various definitions of the term *Business Intelligence* have been proposed, both in academia and industry. Each of these definitions, in their own wording, define *Business Intelligence* as the process by which data is transformed into information which can then be used by decision makers in an informed way. This is done by using various technologies and tools. In this book, we focused on one such technology, namely transactional risk, to achieve Business Intelligence in various business applications. The approaches and techniques detailed in this book systematically tackle the process of assessing transactional risk in business activities and take this into account when making an informed interaction-based decision. Specifically, in this book, we have proposed computational methods for the assessment of the subcategories of transactional risk, according to the context-specific, assessment criteria-specific and dynamic-specific characteristics, combining these assessments to ascertain the severity of transactional risk and evaluate and manage it for the successful completion of the business activity. We believe that the proposed approach for transactional risk analysis in this book, when combined with the existing techniques for trust and security assessments, will assist in providing the required technologies for Business Intelligence in various business applications.

In this chapter, we discuss some of the future research and development in the implementation of the proposed transactional risk analysis approach in various business applications and new research directions.

11.2 Future Research and Development

11.2.1 Knowledge Base of Different Service Providers and Service Users as an Information Repository

As mentioned earlier, business activities are goal specific and therefore, the process of transactional risk analysis, too, is specific to those goals. For the successful completion of this process, service providers and service users need to have relevant information about each other's capability and requirements in the

O.K. Hussain et al.: Risk Assessment & Management in the Networked Economy, SCI 412, pp. 283–287.
springerlink.com

particular context on which to base their decision. Further scenarios may arise where information needs to be solicited from other users (third party or recommending agents) and information about a user (service provider or service user) in relation to certain assessment criteria needs to be inferred from similar criteria due to lack of relevant, pre-existing information in the required criteria. In such scenarios, if each user has their own information repository, then data consistency, uniformity, integrity, completeness and lack of ambiguity cannot be guaranteed. This may result in the user making inferences and decisions based on this misleading information. What is needed is a centralized knowledge base that will act as an information repository to effectively store and share users' information in the different criteria, thereby assisting them in making informed decisions.

11.2.2 Efficient Data Representation, Searching and Knowledge Synthesis Techniques

When a knowledge base of different service providers and users is created as an information repository, on the one hand, it may assist in achieving data uniformity, integrity and completeness, but on the other hand, it may create problems in maintaining its complexity, search efficiency and maintaining a reasonable computation time. This problem is further complicated with an increase in the size of the knowledge base. To address this issue, the efficient representation of data, using techniques such as ontologies and techniques for synthesizing knowledge from data (eg. data mining techniques) are needed.

11.2.3 Multi-criteria Domain-specific Informed Service Selection

As mentioned earlier in this book, transactional risk analysis is one of the important factors in making an informed interaction-based decision. However, it is important to note that it is not the only factor; rather, the assessment of other factors such as trust, security and privacy and their combined considerations are also important to consider when an informed decision on a business activity has to be made. So, a comprehensive multi-criteria service selection approach is needed in different domains such as transportation, digital ecosystems and the cloud computing paradigm etc.

11.2.4 Integration with Existing Systems

To integrate the proposed approach of transactional risk analysis with existing systems, integration functions and related GUIs need to be developed, which will enable the relevant decision makers to better understand the output results, using their pre-existing calculations and analysis.

11.2.5 Maintaining Business Relationships in the Internet Economy

One of the key shortcomings in the existing literature is that the focus is entirely on initiating or building business relationships. Little effort has been made to maintain or sustain previously built relationships. It is important to understand that these relationships have a life cycle, the first stage of which is building them and the subsequent stage is maintaining or sustaining the previously built relationship[1, 2]. The establishment of a business relationship through an adequate level of trust or acceptable degree of risk in such a relationship is neither a simple nor a quick process. It is a lengthy and ongoing process, requiring the concerted effort of both interacting parties [1-5]. Therefore, once a sufficient level of understanding in the business association has been established, and both parties feel that their relationship with each other is valuable; there is then a need to sustain it.

11.3 New Research Directions

In this section, we discuss some research directions in which the proposed transactional risk analysis can be applied in real world business applications.

11.3.1 Service Degradability Management in Cloud Computing

Cloud computing, with its varied features, is an attractive option for both service providers and service users. It provides a service-oriented model to service providers, coupled with features such as multi-tenancy, shared resource pooling, ubiquitous access, dynamic resource provisioning etc. that enables them to maximize the use of their resources. For service users, it eliminates the need to pre-plan and allows the infrastructural resources to be used as services from different vendors at their desired point in time on a demand and pay-as-you-go basis at much lower premiums. To ensure the delivered service meets the desired level of quality, Service Level Agreements (SLAs) between service users and service providers are established. SLAs are multi-faceted agreements that provide a clear understanding of the expected services, thereby eliminating unrealistic expectations and enable planning to take place to optimize the use of resources while maximizing the revenue to be generated. Non-adherence to the SLAs is costly to both the service provider and service user in many ways. For the service provider, it may involve the application of penalties or loss of reputation that may affect the viability of their business; for the service user, it may be the non-achievement of their expectations at the required time which may be critical for their needs and the consequences that flow from this may adversely affect their business.

To avoid such undesirable scenarios, an intelligent framework is needed that in real-time assists both the service providers and service users at different stages of SLAs, from their formulation to their commitment, to ensure compliance. In order to ensure that SLAs are met, approaches have been proposed that utilize an SLA monitoring mechanism [6]. These mechanisms are useful for SLA enforcement and to detect any SLA violation at runtime which may result in *service degradation*. Service degradation is defined as an event that may cause an interruption to the service or a reduction in the quality of the service from what was expected[7]. This, in turn, may result in the non-achievement of desired outcomes, and therefore, should be monitored so that appropriate actions or steps can be taken to mitigate its impact when the chance of service degradation is high. Monitoring service degradation, both from the service user's and service provider's viewpoint, is important. From the service user's point of view, such a monitoring framework will assist him in making an informed decision about forming an SLA with a service provider and continuing with the service, provided it does not go below a certain level of threshold (both in terms of quality and associated consequences). From the service provider's point of view, monitoring service degradation will assist him to achieve the optimal utilization of his resources, while committing to SLAs.

In the literature, although a great amount of work has been done to ensure QoS based on SLAs[8, 9], what is lacking is QoS assessment during actual service delivery. Such techniques can be developed by using the notion of transactional risk discussed in this book. Specifically, by using the proposed techniques, the service requester can determine the transactional risk of a service degrading and utilize this in making an informed service level management decision (whether to continue with the service or recompose the service). The service providers, by using the notion of transactional risk, can in real-time determine the level/s of resource capacity required to commit to a SLA against the level/s of resources available. Such analysis will also represent the level of unused resources which can be used by the provider to form tailor-made SLAs and achieve their optimal use. This is in contrast to the techniques in the literature that dynamically allocate resources to the user's requests [10-13], but in the event of an overload of requests, they switch to different adaptation mechanisms, such as admission control. This may lead in the rejection of services even from users who may have valid SLAs [14].

The application of these techniques will assist in developing a next generation real-time transactional risk analysis-based framework for SLA formation, commitment and assurance that is lacking in the existing literature.

11.3.2 Transactional Risk Analysis in Power Generation Systems

The aim of the problem is to determine the risk and the associated financial consequences to the Genco as the result of the unserved load in the time period. This includes taking into consideration random outages of generating plants and the uncertainties associated with load forecasting. Existing work done in this area simulates the operation of a power system and determines whether the generating plants will generate the required capacity in order to meet the load for a given period of time, which is expressed as Loss of Load Probability (LOLP) and Expected Unserved Energy (EUE) [14-16]. As part of our future work, we intend to extend this analysis and determine the transactional risk of the required load not being generated by the power system in a given period of time.

References

1. Gwebu, K.L., Wang, J., Troutt, M.D.: A Conceptual framework for understanding trust building and maintenance in virtual organizations. Journal of Information Technology Theory and Application 9, 43–63 (2007)
2. Babar, M.A., Verner, J.M., Nguyen, P.T.: Establishing and Maintaining trust in software outsourcing relationship: an empirical investigation. The Journal of System and Software 80, 1438–1449 (2007)
3. Ruppel, C., Underwood-Queen, L., Harrington, S.J.: e-Commerce: The Roles of Trust, Security, and Type of e-Commerce Involvement. e-Service Journal 2, 25–45 (2003)
4. Winch, G., Joyce, P.: Exploring the dynamics of building, and losing, consumer trust in B2C eBusiness. International Journal of Retail & Distribution Management 34, 541–555 (2006)
5. Chang, E., Dillon, T., Hussain, F.K.: Trust and Reputation for Service-Oriented Environments. John Wiley & Sons, Ltd., West Sussex (2006)
6. `http://aws.amazon.com/cloudwatch/`
7. Hammadi, A.M., Hussain, O.K.: Transactional Risk Assessment-Based Approach for Service Degradability Management. In: IEEE International Conference on e-Business Engineering, Beijing, pp. 142–152 (2011)
8. Quillinan, T.B., Clark, K.P., Warnier, M., Brazier, F.M.T., Rana, O.: Negotiation and Monitoring of Service Level Agreements. In: Wieder, P., Yahyapour, R., Ziegler, W. (eds.) Grids and Service-Oriented Architectures for Service Level Agreements, pp. 167–176 (2010)
9. Khader, D., Padget, J., Warnier, M.: Reactive Monitoring of Service Level Agreements. In: Wieder, P., Yahyapour, R., Ziegler, W. (eds.) Grids and Service-Oriented Architectures for Service Level Agreements, pp. 13–22 (2010)
10. Diao, Y., Hellerstein, J.L., Parekh, S.: Using fuzzy control to maximize profits in service level management. IBM Systems Journal 41, 403–420 (2002)
11. Lu, C., Stankovic, J.A., Son, S.H., Tao, G.: Feedback Control Real-Time Scheduling: Framework, Modeling, and Algorithms. Journal of Real-Time Systems 23, 85–126 (2002)
12. Stankovic, J.A., He, T., Abdelzaher, T., Marley, M., Tao, G., Son, S., Lu, C.: Feedback control scheduling in distributed real-time systems. In: Proceedings of the 22nd IEEE Real-Time Systems Symposium, London, pp. 59–70 (2001)
13. Wang, Z., Zhu, X., Singhal, S.: Utilization and SLO-Based Control for Dynamic Sizing of Resource Partitions. In: Schönwälder, J., Serrat, J. (eds.) DSOM 2005. LNCS, vol. 3775, pp. 133–144. Springer, Heidelberg (2005)
14. Yfoulis, C.A., Gounaris, A.: Honoring SLAs on cloud computing services: A control perspective. In: European Control Conference, Hungary, pp. 1–6 (2009)
15. Sanabria, L.A., Dillon, T.S.: Stochastic power flow using cumulants and Von Mises functions. Electrical Power and Energy Systems 8, 47–80 (1986)
16. Sanabria, L.A., Dillon, T.S.: An Error Correction Algorithm for Stochastic Production Costing. IEEE Transactions on Power Systems 3, 94–100 (1988)

MIX
Papier aus verantwortungsvollen Quellen
Paper from responsible sources
FSC® C105338

If you have any concerns about our products,
you can contact us on
ProductSafety@springernature.com

In case Publisher is established outside the EU,
the EU authorized representative is:
Springer Nature Customer Service Center GmbH
Europaplatz 3, 69115 Heidelberg, Germany

Printed by Libri Plureos GmbH
in Hamburg, Germany